TACTICS FOR BASS
AND OTHER
WARMWATER
SPECIES

STACKPOLE FLY FISHING ESSENTIALS

TACTICS FOR BASS
AND OTHER
WARMWATER SPECIES

TIM JACOBS

STACKPOLE
BOOKS

Essex, Connecticut
Blue Ridge Summit, Pennsylvania

STACKPOLE BOOKS

An imprint of The Globe Pequot Publishing Group, Inc.
64 South Main Street
Essex, Connecticut 06426
www.globepequot.com

Distributed by NATIONAL BOOK NETWORK

British Library Cataloguing in Publication Information available

Library of Congress Cataloging-in-Publication Data

Names: Jacobs, Tim (Fly fisherman), author.
Title: Tactics for bass and other warmwater
    species / Tim Jacobs.
Description: Essex, Connecticut : Stackpole Books, [2023] | Includes
    bibliographical references and index. | Summary: "Author Tim Jacobs
covers techniques, rigs, and fly patterns for largemouth and smallmouth
bass, panfish, pike, and other species common in lakes around the
country"— Provided by publisher.
Identifiers: LCCN 2022035764 (print) | LCCN 2022035765 (ebook) | ISBN
    9780811771122 (hardback) | ISBN 9780811771139 (epub)
Subjects: LCSH: Fly fishing. | Warmwater fishing. | Warmwater fishes.
Classification: LCC SH456 .J33 2023 (print) | LCC SH456 (ebook) | DDC
    799.12/4—dc23/eng/20220822
LC record available at https://lccn.loc.gov/2022035764
LC ebook record available at https://lccn.loc.gov/2022035765

CONTENTS

FOREWORD

Whenever I purchase a book on tactics to catch fish, there are a few criteria that I look at before I make the investment. I read the foreword, the introduction, the table of contents, and the bibliography, and this usually indicates the credibility of the author. The bibliography is important because it shows me that he references opinions from other experts and it indicates that this is not an opinion book. Tim Jacobs's book has an extensive bibliography, which shows me he did his research. I then pick a chapter that must answer my questions on a specific topic and be more informative than my present knowledge. When I read this chapter, I can evaluate whether it is presented in an organized way, the ease of reading, and the detail that is expressed. Tim's *Fly Fishing Essentials: Tactics for Bass and Other Warmwater Species* fulfilled all the attributes that make for an excellent "how-to" book.

Catching largemouth bass on topwater bass bugs provides a thrill and satisfaction equal to catching trout on dry flies.

I have known Tim Jacobs for several years, as we tie at the same fly-fishing shows and fly shops. His previous book, *Tying and Fishing Deer Hair Flies*, shows his knowledge of flies that catch warmwater fish. As an Umpqua Fly Designer and a member of the Orvis Pro Team, Whiting Farms Pro Team, and Ross, Abel, and Airflo Pro Team, I understand the necessary attributes that will trigger a fish to strike. For me, knowledge is power, and after reading this book, I realized how little I knew about this subject. A tactical book on warmwater fishing, if followed, bottom line should yield more fish than you are catching now.

A little about Tim's history that makes this book so noteworthy: He started fishing with his grandfather when he was a young boy and has fished for warmwater fish for over 50 years. Being a math teacher for 41 years, he has the natural tendency to be precise, but as an excellent teacher, he has learned to express his knowledge to his students in such a way that they understand the information and retain it. Tim is a fishing guide, FFF- and Sage-certified casting instructor, and Pro Team member for Regal Vises and Nature's Spirit. His book, though technical, is not like reading an encyclopedia but instead is pleasurable reading. I was motivated to continue reading because of the useful knowledge I was gaining that I knew would result in more fish. I have watched Tim tie bass flies with amazement. His precision and attention to detail is tremendous. Most importantly, his knowledge of warmwater fishing contributes to the creation of his fish-catching flies.

Fly Fishing Essentials: Tactics for Bass and Other Warmwater Species gives you a glimpse of the history of warmwater fishing, which always is interesting to see the beginning and how it has evolved. Tim writes about oligotrophic, eutrophic, and mesotrophic lakes and how that affects the success of your fishing day, in addition to the importance of the proper tackle based on what species you are fishing for, the type of water you are fishing, and the type of flies you are casting. This, of course, will prevent you from wasting money purchasing tackle that is not appropriate. Tim addresses the actual process of fishing such as casting, presentation, retrieving the fly, and the approach whether you are using a boat, shore fishing, or wading. He discusses in detail the tactics of fishing for bass, crappie, bluegill, panfish, carp, pike, musky, and walleye as well as the behavior of these fish. Other topics he covers are striking, playing, and landing the fish and the importance of the color of the flies and different fly patterns such as frogs, dry flies, popping, waking, slider, diving, swimming, and jigs.

I have been fishing for warmwater fish for many years and have had success, but the information I have gleaned from this book is eye opening to me and I know it will result in more fish. I would recommend this book to anyone who is interested in a new understanding of the tactics necessary for catching more warmwater fish.

—Phil Iwane

ACKNOWLEDGMENTS

I would first like to thank Phil Iwane for writing the forward for this book. Phil is a friend as well as a gifted fly tier and accomplished angler. His very generous words are much appreciated.

Thanks to all my pro sponsors as well, including Donald Barnes and all the staff at Regal Engineering for their support. Regal manufactures high-quality fly-tying vises that I have been using for many years. I also would like to thank Thomas Lamphere at Nature's Spirit. The deer hair flies pictured in this book would not be possible without the great fly materials that Nature's Spirit provides.

While I am not officially on any of their pro teams, I would like to thank both Sage and Redington for the great rods and reels that I have been using for many years. Additionally, I want to recognize Fishpond for all their great products, especially their innovative tackle bags. Both Scientific Anglers and Rio manufacture great fly lines as well as terminal tackle.

I also want to thank Muskies, Inc., for allowing me to use the pike and musky identification chart.

I want to extend a special thanks to Frank and Janine Whispell for their photographic contributions to this book. Frank and Janine, who are long-time friends, are both talented fly tiers as well as outstanding anglers. You can find out more about Frank at Frank's Fly-Fishing Adventures (www.fffadventures.com).

Finally, I want to thank my family, both in Michigan and here in Colorado, for their love and support. My two daughters, Alyssa and Kelsey, have always brought joy and humor to my life. Most importantly, I want to thank my wife, Susan. She has been both patient and indulgent during the process of my writing this book. As I said in my first book, all that is good in my life has come from meeting and marrying her.

INTRODUCTION

The real voyage of discovery consists not in seeking new lands but seeing with new eyes.

—Marcel Proust
Remembrance of Things Past, 1922

Perhaps less is written about warmwater fly fishing than any other aspect of our sport. Trout, salmon, steelhead, and saltwater fishing garners the most attention from anglers. The irony is that for most of these same anglers, these pursuits require if not a destination trip, then at least a long drive. Likely a warmwater resource is nearby, and these overlooked fisheries often offer good fishing. While destination trips and travel are wonderful, daily fishing is what hones an angler's skills. Additionally, with many trout fisheries warming up during the summer, conservation groups and state wildlife agencies are requesting that anglers avoid trout fishing when rivers warm up to 70 degrees Fahrenheit or higher. The Anglers of the Au Sable in Michigan ask anglers to take the "Seventy Degree Pledge" and refrain from trout fishing when the river reaches that temperature. So here, many anglers have a smallmouth bass or other backup fishing plan during the summer when waters reach these temperatures.

For the saltwater destination angler, fishing lakes for bass and other species serves to hone the casting skills needed for the saltwater flats. This is especially true when fishing out of a boat. Most warmwater lake fishing requires the ability to cast at least 50 to 60 feet. Using larger-weight rods and bigger flies requires the use of a double haul cast. I did not feel out of

This wild lake in the northern Greats Lakes provides exciting fly-fishing opportunities.

place the first time I was on the bow of a flats boat in Florida Bay. We were fishing for redfish, and while I did need to make somewhat longer casts and deal with more wind, the warmwater lake fishing helped to make the experience one that was somewhat familiar.

While warmwater fly fishing can serve as an alternative when coldwater fisheries become too warm or help you prepare for saltwater fishing, it is a fly-fishing experience deserving of attention for its own merits. The tackle required along with the unique and artistic nature of many of the flies should make it a prominent component of any angler's experience.

My own journey to making warmwater fly fishing a salient part of each season is like many other anglers' experiences. As chronicled in my first book, *Tying and Fishing Deer Hair Flies*, I started fishing as a young boy with my grandfather. We fished from our cottage on Lake George in northern Michigan. As with many of the lakes in this region, it was primarily a warmwater lake with largemouth bass, pike, and numerous species of panfish. When I first started to learn to fish with a fly rod, I was definitely focused on trout fishing and would use our cottage as a jumping-off point to fish the Au Sable and Pere Marquette Rivers. Like many other anglers, I was guilty of overlooking a fishery literally right under my nose. Over the course of my fishing career, I have driven by many good fishing opportunities to travel to well-known rivers or lakes. For many years I was ignorant of a superb smallmouth fishery that was only a few miles from where I grew up. When I finally began to look around with a different mindset, I realized that I had been missing a lot of good fishing nearby.

As a guide in Steamboat Springs and the Roaring Fork Valley in Colorado, my fishing days were spent pursuing trout. Trout fishing usually means that you are fishing someplace that is primarily a trout fishery and you are catching primarily trout. Warmwater fishing is rarely fishing for just one species. These fish coexist with each other in most waters, so a typical day means that several species of fish are normally caught. Barry Reynolds in his book *Beyond Trout* recounts a story of fishing some warmwater ponds in Colorado and catching seven different species in one outing. This is one of the unique and special aspects of fly fishing for warmwater fish.

Warmwater fly fishing does require a different approach from trout fishing. Trout tackle and flies will work for bluegill and other panfish; however, casting the larger flies used for bass, pike, and musky requires different tackle, flies, and tactics. Over time I discovered the proper tackle, tactics, and flies that work for these fish. This book is modeled after the idea that a fishing outing will encompass catching numerous species of fish, and so it includes an overview of the tackle, flies, and tactics for fishing these warmwater species. There are other books that delve into greater detail about a single species; they were good sources of information as I began to pursue a particular warmwater species of fish and would serve you as well. I also did not illustrate how to tie the knots that I reference in the text. It is easy to find information on tying any of the knots mentioned here on the internet or in other books.

I wanted to write a book that was more representational of my fishing days. I have spent the most time fishing for largemouth bass and panfish; however, I have enjoyed fishing for all of the species included here.

If you haven't done so yet, find some warmwater fly-fishing opportunities near you and, perhaps like me, it will become one of your favorite ways of using a fly rod.

1

A Brief History of Warmwater Fly Fishing

The Angler must intice, not command his reward,
and that which is worthy millions to his contentment,
another may have for a groate in the Market.

—Gervase Markham, *A Discourse of the*
Generall Art of Fishing (1614)

The history of fly fishing dates back to antiquity, and our ability to discern its true origins is difficult. Early depictions of fishing, with what appears to be a fly, date as far back as Bronze Age civilizations. From that early time, angling with a fly can be traced through to the Middle Ages in Europe. Some of the early British works are thought to have been influenced by the early French angling oeuvre.

Perhaps one of the earliest folios is *The Treatyse of Fysshynge wyth an Angle* (1486) by Dame (or Lady) Juliana Berners, a prioress of the nunnery at the Abbey of Saint Albans (which is today the Cathedral and Abbey Church of Saint Alban, in St. Albans, a small town north of London). The *Treatyse* begins with a discourse on why angling is superior to all of the other sporting endeavors of the day, then proceeds to give instructions

Topwater deer hair bass bugs are as effective today as they were when anglers first began using them for largemouth bass.

on how to construct tackle and descriptions of fishing methods and a dozen flies. What the *Treatyse* makes clear is that fishing was done with a rod, line, and hook. Today we take for granted that definition of fishing; however, six centuries ago this definition was not necessarily the common thought. This concept separated it from other forms of fishing and elevated it to the status of hunting, which at the time was considered the aristocratic and premier outdoor sport. The *Treatyse* existed for a century and a half as the most prominent work of angling literature.

Izaak Walton's *The Compleat Angler*, first published in 1653, is second only to the *King James Bible* as the most reprinted book in the English language. Like Berners, Walton commends the virtues of angling, using a discourse between an angler, hunter, and falconer. Once "Piscator" has convinced the others as to the superiority of angling, Walton continues to instruct as to the making of tackle and advise on angling tactics for a wide range of fish species. By this time, it was assumed that angling meant to fish with a rod, line, and hook. Flies are mentioned, but a generous description of bait fishing is also included. It wasn't until the fifth printing (1667) that Charles Cotton's section on fly-fishing instruction was added, securing *The Compleat Angler*'s place in fly-fishing history. With Cotton's addition, the genesis of modern fly fishing began.

In subsequent angling literature throughout the next two centuries, most of the writing revolved around trout and salmon. It was when these traditions were exported to North America that the idea of fly fishing expanded to include other fishes.

Theodore Gordon, considered the father of American dry-fly fishing, also popularized streamer fishing with his Bubble Puppy fly pattern. Gordon began tying and experimenting with streamer fishing in the late 1800s, fishing not just for trout but for a variety of fish, including pike, striped bass, and other warmwater fishes. Unfortunately, Gordon never wrote a book, so accounts of his thoughts on fishing have had to rely on letters he wrote along with the writings of many of the contemporary anglers at the time.

The *Treatyse* established that fishing was done with a rod, line, and hook.

Book of the Black Bass by Dr. James Henshall, published in 1881, is one of the first works focused on bass. While it included a description of fishing a "bob" (which was basically some deer hair, perhaps a tail, attached to a hook) for bass, the fly patterns were simply larger versions of the wet flies of the time that were used for trout and salmon. However, Henshall is the most influential writer of the 19th century for expanding our understanding of fly fishing as a method for more than trout. He unequivocally stated that bass were "inch for inch and pound for pound, the gamest fish that swims."

Mary Orvis Marbury's *Favorite Flies and Their Histories* (1892) included 10 plates of bass flies. They were all enlarged and perhaps more heavily dressed versions of the wet flies of that era. This typified the patterns that were being fished for bass at the time.

Black Bass, Florida, by Winslow Homer. Watercolor over graphite, 1904.

The American artist Winslow Homer is noted not only as a pioneering watercolor artist of the late 1800s but also is remembered for his many depictions of fly fishing. Much of Homer's early work was inspired by his trips from his home in Maine to the Adirondacks and Quebec. With photography still in its infancy at the time, Homer's art is a window into the outdoor life and angling for the native brook trout that existed in the Northeast during the late 19th century. It was not until his 50th birthday, in 1886, that he traveled to Florida and began fly fishing for largemouth bass. Numerous paintings during this period depict the bass fishing that existed in the St. Johns and Homosassa Rivers. The paintings clearly show bass being hooked with flies that closely resemble the large wet flies used during this period.

It was not until the early 20th century that warmwater fly fishing began to break away from its British roots. One of the earliest flies that broke with the standard wet flies being used was Orley Tuttle's Devil Bug. It was an all-deer-hair fly tied to represent the large beetles that the smallmouth bass were eating as they hit the water in the lakes surrounding his home in the

Fly rods were commonly used for bass and bluegill up until the end of World War II. Note the fly rod in my grandfather's hand in this photo. However, much like in Izaak Walton's time, this rod was not used for artificial flies only.

Adirondacks. After his first ties in 1919, Tuttle began to market them and by 1922 was selling 50,000 bugs a year.

Ernst Peckinpaugh of Tennessee was also designing and selling bass poppers beginning in the 1920s. His poppers are the precursor of many of today's bass bugs, consisting of a cork body followed by a hackle collar and feather tail. This basic design has endured for a century.

Another influential fly tier and guide during the mid-1900s was Joe Messinger. His Messinger Frog was one of the early examples of flared or spun deer hair that has become a standard modern fly-tying technique for large patterns. While Messinger employed a different technique to flare deer hair than that used by most modern deer hair tiers, it is still an example of exemplary deer hair work. Perhaps this fly's greatest contribution is the way in which it was engineered to be fished effectively. The deer hair body was rounded, with hair on the bottom of the hook shank as well as on the top. As the body of the fly soaked up water, this underbelly, along with the weight of the bend of the hook, provided the balance needed so that the fly would land upright in the water every time it was cast. Many modern fly tiers, myself included, have adopted this concept in their deer hair bass bugs.

Prior to World War II, bass were commonly fished with a fly rod, using flies designed by Tuttle, Peckinpaugh, and others. The revolving spool bait-casting reels of the day required an educated thumb to prevent the reel from over-spooling and tangling the line, so they were not easy to use—a lesson I learned early as I tried to fish with my grandfather's rods. The introduction of the modern spinning reel contributed to the decline of warmwater fly fishing, especially for bass. The spinning reel's ease of casting made it popular, especially with casual anglers. At least from a literary standpoint, fly fishing was more the realm of trout fishermen during the late 1950s and into the early 1970s.

Two works of this period that continue to stand the test of time are Ray Bergman's *Fresh Water Bass* and Sid Gordon's *How to Fish from Top to Bottom*. While neither book is directed exclusively at fly fishing, both include the use of fly rods. Bergman's book does for largemouth bass fishing what his *Trout* did for trout fishing. In typical Bergman style, it includes a lot of information about bass fishing, along with a generous helping of anecdotal stories to help illustrate his theories. Sid Gordon was a conservation officer for the state of Wisconsin. His advice regarding lakes and fishing focuses on the Great Lakes region, and his understanding of the lakes in this area is similar to my own experience.

Like much of the writing of this and earlier periods, Bergman's and Gordon's information needs to be translated to account for modern tackle and fly patterns. However, fish and their habitats have a certain constancy, and both books are as relevant today as when they were written.

A. D. Livingston's book *Fly Rodding for Bass* appeared in 1976. Like most of the fishing literature that came before, in it Livingston covered tackle, tactics, and flies for bass. However, it was Dave Whitlock who brought fly fishing for bass into the consciousness of many anglers in the 1980s. His *Fly Fishing for Bass Handbook* became the go-to guide for many who were first venturing into the warmwater sphere. I know that my 1988 dog-eared copy was a constant resource when I first began to pursue bass with a fly rod.

Today virtually every species of fish that swims in freshwater is pursued with a fly rod. Additionally, during the last three decades all of the species of fish included here, with one exception, have had books dedicated to them. Previous generations of fly fishers would be amazed by the abundance and availability of fly-fishing knowledge that is so readily available today.

2

Understanding Lakes

*When the line is unrolling in a long, sweeping curl like
the crest of a comber breaking on a smooth stretch of
beach . . . physics and art united in graceful symmetry
. . . it is easy to believe that fly casting is an end unto
itself. Inevitably, though, . . . the fly at the end of your
leader causes the idea of interacting with a fish to make
its way onto your list of intentions.*

— Cliff Hauptman, *The Fly Fisher's Guide to Warmwater Lakes*

The expectation of pulling up to a boat ramp on a
new lake is one of those joys that continues to pull
me back to warmwater fly fishing. On the other side
of this is returning to familiar lakes that I have fished
before. Some I have fished for decades, while others
perhaps only a few times. Either way, all of these places
are part of the tapestry woven from my devotion to
warmwater fly fishing.

The Great Lakes region, and particularly my native
state of Michigan, has a great number of these warmwater lakes. It would take a lifetime of fishing to even
begin to make a dent in all the available lakes. Many
of these underutilized fisheries provide an excellent
opportunity for the fly fisher. While most of these
lakes have many common features, they are all also
unique, with none of them being exactly like another.

The Great Lakes region has an abundance of small lakes
like this that have a diverse population of warmwater
species of fish.

Warmwater lakes are so numerous across the Great Lakes region that even conventional-tackle anglers do not put an undue amount of pressure on them, and they seem to be completely overlooked by fly fishers.

On Lake George, my home lake, most anglers are trolling in the deeper parts of the lake, with a few anglers using conventional tackle casting to the shoreline. In all my years here, I am really the only one who seriously fishes with a fly rod. These fish just do not see flies, so when they do, they rarely hesitate to take them. I know some avid Michigan fly fishermen who actually live on warmwater lakes and yet never uncase their fly rods until they get to one of the better-known trout streams. With a small boat, suitable rods and lines, and the proper flies, anglers can find many angling opportunities close to home.

When searching out suitable lakes, it can help to understand the different types of lakes and focus on those that are best suited to fly fishing. Geologically, lakes fall into three categories: oligotrophic lakes, eutrophic lakes, and mesotrophic lakes.

OLIGOTROPHIC AND EUTROPHIC LAKES

Oligotrophic lakes are those that are considered geologically younger and are deeper and colder than the other two types of lakes. They normally support a population of lake trout and perhaps other trout. There will be little to no vegetation along the shorelines, and the water in these lakes will normally

Oligotrophic lakes are some of the most beautiful but rarely offer fly-fishing opportunities.

Eutrophic lakes are often too shallow and filled with weeds for fly fishing.

be clear. In Michigan, Glen Lake and Torch Lake, with their turquoise-blue waters, are considered two of the most beautiful lakes in the state. Both are considered oligotrophic. These lakes are best left to the anglers who are trolling with outriggers or using other methods for fishing in deep water.

Eutrophic, and especially hypereutrophic, lakes are at the other end of the spectrum. These lakes are typically filled with weeds and tend to be shallow. Often they will have algae blooms in the summer. The bottoms of these lakes will be soft and largely composed of muck formed from decaying vegetation. They will support few fish, as they tend to be very warm in the summer and the decaying vegetation uses most of the dissolved oxygen in the water, leaving high levels of carbon dioxide for the fish that are there. These lakes are difficult to fish and at the extreme level not worth the effort to do so.

MESOTROPHIC LAKES

Mesotrophic lakes (or lakes that are slightly eutrophic) are the sweet spot for the fly fisher. As lakes geologically age, they move along the scale from being oligotrophic to eutrophic. Since this aging takes place along a geological timeline, it takes a very long time to happen, but as it does, they move

Medium to small mesotrophic lakes, with their water clarity, shoreline cover, and structure, offer some of the best fly-fishing opportunities.

through the mesotrophic stage. These lakes tend to be of medium depth, with some shoreline vegetation. They should have some hard, sandy beach areas (this does not imply a swimming beach) for at least some of the shoreline. These hard-bottomed areas are where bass and panfish will spawn. The lighter circles that are present in the spring and early part of the summer are the spawning nests that different species of warmwater fish have created.

These lakes also tend to have clear water (though not as clear as oligotrophic lake water), although in some areas the lakes can take on a tannic or tea color if there is an influx from an inlet where a cedar swamp is present. Often as you move out into the lake from these beach areas there are drop-offs into deeper water where weed lines are present. Points are another feature to look for. Many times a point will have a hard beach area around it and then a drop-off into deeper water. Points can be fish magnets, as fish can easily move around the point as conditions change during the day.

Another feature of many of these lakes is a shallow, muck-bottomed bay. The bay will have lily pads, with reeds and cattails lining the shoreline. This area of the lake is where the water first begins to warm up in the spring. The bay itself can be difficult to fish if it is too shallow; however, the entrance to these bays can provide good fishing. Deeper bays that have a consistent depth with weed growth that does not reach the surface of the water can also provide good fishing. Fish use the weeds for shelter and cover while at the same time keeping a watch on the water just above the weeds for a feeding opportunity.

Bays with weeds, both underwater and on the surface, are prime targets for the fly fisher.

Lily pad beds provide overhead cover as well as edges. Fishing the irregular edge is a productive strategy.

Islands or mid-lake humps also provide fish-holding areas. These islands offer an additional shoreline, and many of the same features that exist on the main shoreline exist here as well. Mid-lake humps or rock piles also provide a structural feature that holds fish. If a lake has an inlet or an outlet, these may be areas to fish as well. Especially if there is an inflow of water that is moving, fish can settle off to the side of these currents and intercept any food items that may come drifting by.

Lakes also need cover for the fish. Fish relate to cover for a variety of reasons: as an ambush point for intercepting food, a hiding place, or a place to find shade. Cover can come in many forms. On midwestern and eastern lakes, cover normally includes beds of lily pads. Lily pads do not uniformly cover the water but will have notches that extend back into the bed as well as channels that go through the bed. While fish do reside in the shade under the pads, they are very aware of the edges. Most fly fishing is done on the edge features, with casts up into the notches or back along the channels. If open water is present on the shoreline side of a bed of lily pads and you can maneuver into position, these areas are worth a cast. Fish can be present on the shallow-water side of the lily pad beds as well as the deep-water side.

Other forms of natural cover can include reeds, downed trees along the shore, and other types of vegetation. On developed lakes man-made structures play a role similar to that of natural cover on undeveloped lakes. Swim platforms, docks, boathouses, and anchored boats (especially pontoon boats) all provide shade and cover for fish. Fish can use this man-made cover as both an ambush point and an area shaded from the sun.

Canals on lakes are another feature that can hold a surprising number of fish, depending on how much boat activity occurs there. When fishing a canal, be sure to fish the entrance first. A quiet approach as you work through the canal is also essential, as any sound or disturbance is amplified in this small environment. Be sure that you have the water depth to enter the canal. Sometimes decayed vegetation can create very shallow water. I know from personal experience that is it difficult to extract yourself from a canal that suddenly becomes too shallow for your boat.

LIGHT

Light plays a critical role in where fish will hold. Most warmwater fish will choose to stay in shaded areas when given the chance. The common wisdom for fishing warmwater lakes is that the fishing is best early in the morning and late in the evening. Fishing can be good during those hours because there is a lot of shade at those times. However, warmwater fishes are sight feeders, so will be actively feeding during the daylight hours as well as the low-light hours. When fishing the midday hours, continue to look for shade created by docks, lily pads, or other cover. Although bass and other predatory fish do like shade, I have occasionally caught them out in brightly lit areas during the day.

Fish will often hold in a shaded area during the morning hours.

EDGES

When trout fishing in rivers, current edges are the main attribute you look for to find fish. In places where faster water and slower water meet, trout will often sit in the slower water, using the faster water as a conveyer belt to bring food items to them. On land, edges are the places where the most wildlife reside. Game birds, deer, and other wildlife use the areas where the woods meet a meadow or field for protection and finding food. Similarly, edges become the most important feature when locating warmwater fish.

There are a variety of edges in lakes. The most important and most common one is where the water meets the shore. On many developed lakes, landowners have constructed seawalls along the front of their property. Many times there is some water depth along the wall. Be it from six inches to a couple of feet of water, this becomes a good barrier for predator fish to capture prey fish. Bass and other predatory fish will sit along these seawalls, where they are in the shade, and wait for an opportunity to feed. Lakes that have undeveloped shoreline will have other features such as downed trees and vegetation that create edges as well.

In addition to providing cover, docks, boats (especially pontoon boats), and other man-made structures also create edges. Fish will rest in the shade below a dock or boat, waiting to ambush a smaller fish that swims past. If the sun is directly overhead either side can be fished, but normally the sun will be creating more shade on one side of the dock or boat. Where the shaded area meets the sunny area is also an edge. Fishing a fly along the side in the shaded area is often productive. One caution, however: boat owners do not like you hitting their boat with your fly or fly line.

Fishing edges and cover has always been productive. My great-uncle and great-aunt, Ralph and Nora Harrison, are fishing the edge of the reeds on Lake George, circa 1950.

Shorelines that have hard sand bottoms, cover, and structure and taper off into deeper water can be fished with a variety of methods.

Other edges include the edges of lily pad beds or reeds, and they also exist along lines of underwater weeds. Trees that have fallen into a lake also provide both shade and edges for fish to utilize. Another, more ephemeral edge is simply a shade edge in open water along a shoreline with large trees. Early and late in the day the trees will provide shaded areas that extend into the lake. Fish will often cruise these shaded areas of a shoreline, looking for a feeding opportunity. Fishing from a shaded area also gives you the advantage of not being seen by the fish.

TEMPERATURE

Temperature is another factor when considering fish behavior. All of the fish discussed in this book have their own unique temperatures that they prefer; however, the temperatures where they are most active are somewhat similar. Typically, the fishing starts to get good when the temperature of the lake begins to hit and exceed 70 degrees. Generally speaking, the fishing is better as the lake temperatures are rising, while it slows down if a cold front moves through, making the lake temperatures drop. If temperatures get too hot, exceeding 85 degrees, the daytime fishing can sometimes slow down. When this happens, morning fishing will usually be more productive. This may be truer of northern lakes than southern lakes.

PONDS

It is difficult to say what the difference is between a small lake and a pond. Shingle Lake, in Michigan, sits next door to Lake George. At just 35 acres, Shingle Lake feels small. I will typically

Ponds provide good warmwater fishing opportunities in western states.

PHOTO BY FRANK WHISPELL AT WWW.FFFADVENTURES.COM

Fly fishing in ponds can be productive, if there is enough room for your backcast.

circumnavigate the lake several times in a fishing outing there. It deserves the name of lake, but it is bordering on being a pond.

Ponds will have the same kinds of cover and structure that lakes have in whatever region of the country they are in. In Colorado, ponds do not have lily pads but do have reeds and cattails along the banks. Ponds in the Midwest will have all the same kinds of vegetation that the lakes have. One of the differences in ponds is that if the pond is privately owned, the makeup of fish might be different from that of the surrounding lakes. All the same things that affect the fish in lakes will affect them in ponds. The one difference is that if the pond is small enough, you can be certain that the fish are seeing your fly as you retrieve it, and they can readily see you as well.

MAN-MADE LAKES

Man-made lakes include both small lakes that have been enlarged by damming an outlet and reservoirs created by damming a river. These lakes and reservoirs can be looked at the same way that natural lakes are looked at, with one exception. These bodies of water will often have old tree stumps in them. In fact, if you see stumps in a lake, there is probably some form of dam, though small, that raised the level of the water. Stump fields provide great cover for fish. However, this is the place where the heaviest leaders need to be used, as you must get the fish away from the wooded structure as quickly as possible.

SEARCHING OUT THE RIGHT LAKES

Lake maps are an indispensable resource for finding lakes that are likely candidates for exploration. I like to fish smaller lakes that are from 30 acres to perhaps 400 acres. With my 14-foot boot and small outboard, if I venture into lakes that are much larger than that, travel time can become an issue. If the map of the lake has numerous contour lines close together near the shoreline, that means the shore drops off quickly into deep water. If this steep drop-off is present all around the lake, that lake will not normally be a good candidate for fly fishing.

Look instead for lakes with shallow-water shelves near drop-offs. The drop-off does not have to be great—even a few feet will do. On a lake map the contour lines will be spread out, indicating shallow flats. On the lake, if these shallow-water areas have some cover in the form of lily pads, reeds, or man-made structures, the chances are good that bass and other warmwater fishes will be present.

The amount of development on a lake is also a consideration. Development doesn't necessarily mean that the fishing is bad. Lake George (my home lake) has cottages and homes around most of the shoreline, but it is an excellent bass lake. On these types of lakes, instead of casting to natural cover, I cast primarily to boat docks, swim platforms, and other man-made structures. It is not always easy to tell the amount of development from lake maps.

You can also use Google Maps to check out what a lake looks like. Along with seeing how developed the lake is, you are often able to find the access

The presence of stumps indicates that at some point the water levels of this lake were lower than those today. There is a lot of structure and cover here, making this a prime shore for fly fishing.

The open water along with the reeds and lily pads make this a prime lake for flies.

point and perhaps even see the boat ramp. I use both Google Maps and lake maps to research any new lakes that I am thinking of fishing. Lake maps can be a hard copy, as I have numerous books of lake maps, though many of the same maps are online today.

A SAMPLING OF LAKES

One of my favorite lakes is Lily Lake in Clare County, Michigan. This 161-acre lake has an abundance of shallow-water cover for the fly fisher. It would probably be classified as slightly eutrophic, since is it pretty shallow with a lot of weed growth, but it is easily accessed with a boat. With a large amount of undeveloped shoreline, there is lots of natural cover in the form of lily pads and reeds. The greatest depth in the lake is about 14 feet. The reed islands directly out from the boat ramp are a good place to start. This lake tends to

be a good topwater bass bug lake, as most of the cover is from 1 to 4 feet deep. If swimming flies are used, they can normally be fished with just a floating line and leader.

Arbutus Lake in southeast Grand Traverse County is another good lake in Michigan for the fly fisher to try. This scenic 395-acre body of water is divided into five different lakes, with passages of varying sizes between each lake. It probably would be classified as mesotrophic, as it has a lot of sandy hard-bottom shoreline and numerous points. The public boat launch is on lake #2. It is a good concrete boat launch with plenty of parking. A good place to begin fishing is the shallow bay on lake #2 across from the boat launch. Try the edges of the lily pad cover early and late in the day with topwater bugs or diving deer hair frogs. During the middle of the day, try the deeper areas of this bay with a sink-tip line and Woolly Buggers or other swimming flies. Lakes #1 and #5 are also worth exploring.

One of my favorite western lakes is Pelican Lake in Utah. This lake is slightly larger than 1,000 acres and is listed as one of the state's Blue Ribbon fisheries because of the trophy bluegill that can be caught here. These bluegill are big and can easily put a deep bend into a 7-weight rod. While it can be fished from a float tube, having a boat with an outboard will open up the entire lake to be fished. However, one caution on this high desert lake: it will have a boom-and-bust cycle and will occasionally winterkill. It is definitely a good idea to get a fishing report before traveling here.

These lakes just scratch the surface, providing only a small example of the variety of nearly unlimited lakes that are available. If fishing in the Northeast or places like the Canadian Shield lakes in northern Minnesota, then smallmouth bass will become the more dominant species in most lakes. These lakes tend to have colder water temperatures, and most have rocky bottoms that smallmouth bass favor. If fishing in the southern United States, then large reservoirs are more the norm for largemouth bass. Southern largemouth bass can take advantage of longer growing seasons and are larger than the largemouth bass in northern lakes. Wherever you are fishing and whatever size lake you are fishing, all of the principles of locating and catching bass still apply.

3

Modern Tackle for Warmwater Fly Fishing

I doubt if rifle, shot-gun, or fowling-piece ever becomes
so dear and near to the sportsman as the rod to
the angler.

—Dr. James A. Henshall, *Book of the Black Bass*

Warmwater fly fishing does require a different approach when compared to trout fishing. Trout tackle will work for fly fishing for bluegill and other panfish; however, casting the larger flies used for bass, pike, and musky fishing requires different tackle.

FLY RODS

When I first began to fish for bass and other warmwater fish, the only rods I owned were a 4-weight rod and 6-weight rod that I used for trout fishing. Both rods were fiberglass and neither one was very useful for casting a large bass bug. I then got a fiberglass blank from Art Neumann's Wanigas fly shop in Saginaw, Michigan. It was a 9-foot rod for an 8-weight line. At the time during the 1980s there was some thought that slower, softer fly rods were needed when fly fishing for bass. It was theorized that these rods were better suited to cast large, wind-resistant bass

Having several rods rigged and ready to fish will facilitate quickly taking advantage of evolving fishing opportunities.

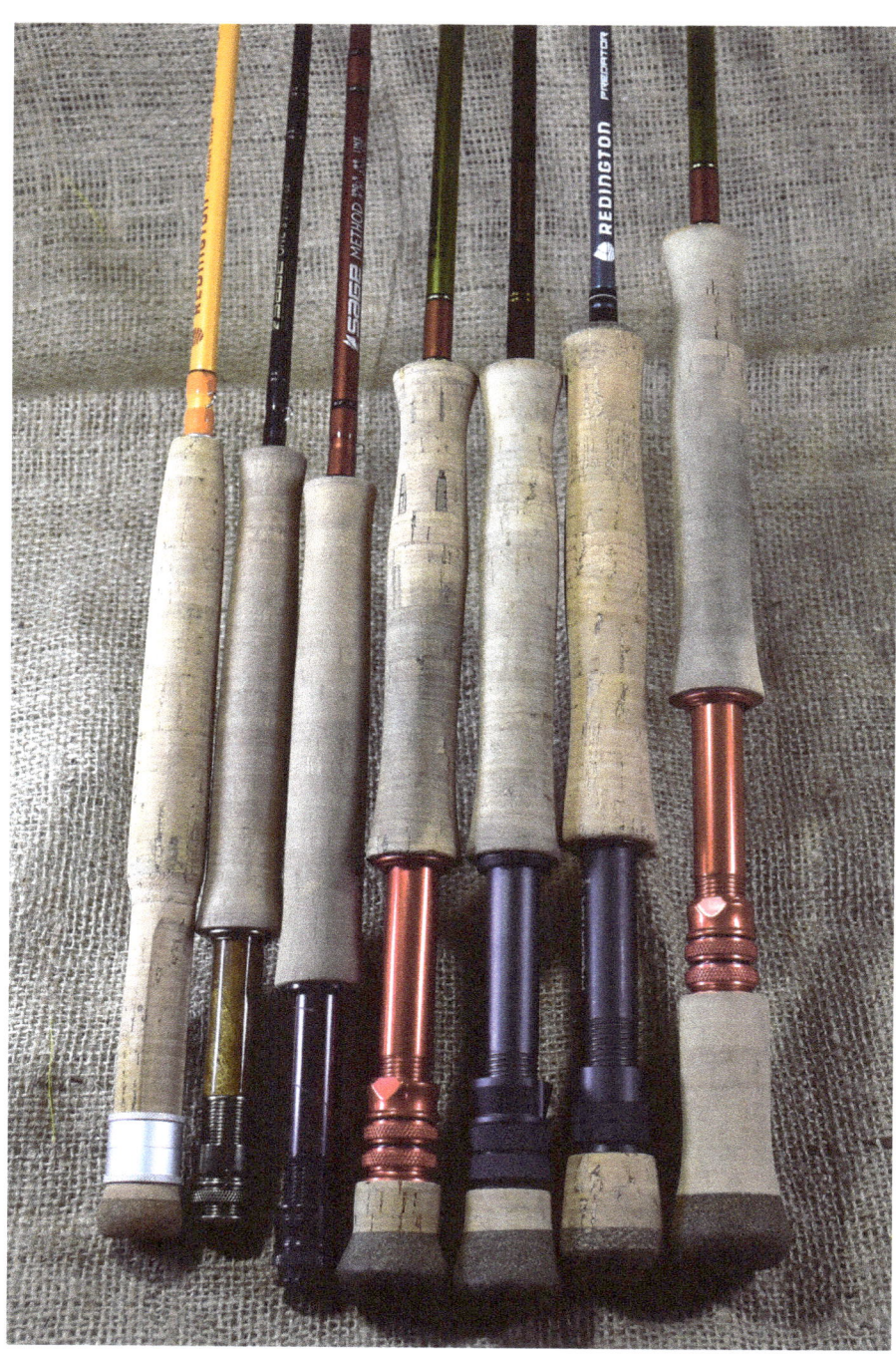

A selection of fly rods ranging from a 2-weight up to an 11-weight can be used for warmwater fly fishing. However, two rods can get the job done. A 4-weight rod will work for all bluegill fishing, and an 8-weight rod can work for all the other fishing.

bugs because the wind resistance of these larger flies tended to slow down the line speed, so a slower rod was needed.

When a rod is classified as fast or slow, what this is really referring to is how stiff the rod is or how resistant it is to bending. It also is measured by where the bend in the rod occurs when casting a set length of line. The faster rods will bend more near the tip as the rod is loading to cast the line. A softer rod will bend more down into the butt of the rod with the same amount of line. A stiffer rod will propel the line with a higher speed than will a softer rod; consequently, the label of fast or slow has been applied to describe a rod's action.

One of my first graphite bass rods was a Scott that was 9 feet long and had two tips. The first tip had standard snake guides and was rated for a 6-weight line. The second tip was rated for a 7-weight line and had single-foot ceramic guides. This made the rod a dual-purpose rod for trout and bass. The single-foot ceramic guides were thought to slow down the rod slightly and be more efficient at shooting bass taper lines. Compared to my fiberglass rod, this rod was an improvement largely because it had a faster action than the fiberglass rod. When I got my first Sage RPL 9-foot, 8-weight rod, I realized that the slower-rod theory was not true. This Sage rod was the fastest rod I had used to that date, and it was definitely the best of the three rods for casting a bass bug taper line and larger bass flies. Fly rods for warmwater fly fishing kept getting better with time, and as with many things, they became more and more specialized.

Today many rod companies are producing specialized rods for warmwater fishing. I have been using primarily Sage Bass II rods for most of my warmwater fishing in recent years. These rods are 7 feet, 11 inches long. There are some considerations needed to successfully cast these shorter rods (see the chapter 4 sidebar "Casting Tournament-Legal Bass Rods"). Other fly tackle companies also produce shorter rods for bass fishing. These rods are normally less than 8 feet long.

With the shorter rods that are less than 8 feet as compared to rods that are up to 9 feet long there are tradeoffs. The shorter rods are more challenging to cast, as you give up some of the leverage on the line; however, what is given up with casting is gained back when hooking and fighting a fish. With a rod that is closer or equal to 9 feet long, casting becomes easier, but you have to pay attention to how you strike a larger fish as well as how you fight it.

When it comes to considering rods for panfish, most of your normal trout rods will work fine. Rods that are rated for a 2-weight line up to a 5-weight are all suitable for panfish, provided the rod used matches the fishing conditions. The lighter 2-weight or 3-weight rods become difficult to cast if is too windy; however, under the right conditions lighter rods can be a lot of fun to use for panfish. While a 6-weight rod might be considered heavy for panfish, it has several advantages. A 6-weight will handle small deer hair bass bugs better than the lighter-weight rods will. Also, a sink tip can be added to fish deeper depths, and this rod will do well casting the added weight of the line plus tip. The Sage II Bluegill rod and the new Sage Payload 6+ have been my standard rods for panfish. Both cast slighter larger flies with ease as well as handle casting an added sink tip. A standard 9-foot, 6-weight rod will work well with the larger flies as well.

One of my favorite rods in recent years is the Sage Bass II Smallmouth rod. It was designed to cast a 290-grain line and is generally considered an 8-weight rod. This rod easily handles an added sink tip and can cast larger bass bugs (e.g., a bass bug tied on a TMC 8089 size 2 hook) with ease. The Sage Bass II Largemouth rod is considered more in the 10-weight range. Ten-weight rods can be used in certain situations, but for bass fishing, an 8-weight rod is more suitable for most conditions.

When fishing for pike or musky, normally larger rods in the 10-weight or 11-weight range are preferable. Sage produces both the Pike rod, which is a 9-foot 10-weight, and the Musky rod, which is a 9-foot 11-weight. The Payload rods are the newest line of rods from Sage that are designed to cast larger flies. Other companies also produce specialty rods. Under most circumstances, pike and musky fishing does call for these larger rods. If you spend enough time fishing for these species, a specialized rod is definitely worth the investment. However, if you only occasionally pursue these fish, use your standard 9-foot rods of the appropriate weight.

FLY REELS

Small-arbor reels were the norm when I first began to fly fish. Today it is difficult to find them, as most companies have completely changed their offerings to large-arbor reels only. Large-arbor reels are vastly superior to the older smaller-arbor reels. A large arbor can quickly reel up slack line when either playing a fish or reeling in line when getting ready to move to a different location. For a given amount of slack, a small-arbor reel can take up to four or five times as many revolutions of the reel to pick up the same amount of line as a single revolution with a large-arbor reel. Additionally, the line comes off the reel in larger coils, so stretching the line to straighten it is easier. With the older reels, I often would have to take a few moments to stretch and straighten the fly line before beginning to cast. When playing a fish, the drag is more consistent, since the diameter of the reel remains consistent where the line is spooling off.

The drag systems on most large-arbor reels are self-contained and do not need any maintenance. After a few fishing outings, I will pull off the spool and wipe the reel using a soft cloth with a small amount of Armor All applied to it. This is about all that is needed to maintain the reel. At the end of the

Almost all the reels I use today are large arbor, as they perform better in all fishing situations.

season during the winter, I will unspool the line and backing and do the same wipe-down of both the spool and frame as well as clean the line with one of the cleaning products manufactured by the line companies. Always store the reel with the drag backed off as far as it can go.

DACRON BACKING

When setting up a reel, I use 30-pound-test Dacron backing for all my rods that are 6-weight and larger. I begin by attaching the Dacron using a uni-knot, also known as a Duncan loop knot. This is a slipknot so that it can be tightened up to the spool. However, one loop of backing does not provide enough friction, so it will slip around the spool when you attempt to reel it in. Because of this, I add several whip-finish knots over the Duncan loop. Once I have 8 to 10 loops of line around the spool, I test the backing to see if it will spin on the spool if I try to reel it in under tension. Normally these extra loops provide enough friction with the spool that the backing does not slip. Most reels are designed to hold 100 yards of backing. I will normally start there and remove some of the backing if the fly line overfills the spool.

At the end of the backing where the fly line attaches, I used to do a dou-bled-up surgeon's loop. This is a simple knot, and by having a double loop with two strands of the backing to attach to the fly line's welded loop, you

This reel is loaded with 30-pound-test Dacron backing. A reverse splice loop has been "tied" at the tip of the backing. Note that this knot is not really tied as much as spliced. The reverse splice loop needs a fine wire to thread the backing through itself. I used 0.009-inch nickel E string for a guitar. These strings are relatively inexpensive and do the job nicely in both 30-pound-test and 20-pound-test Dacron backing. Most modern fly lines have welded loops at each end of the line. With a loop at the end of the backing, a fly line can be easily attached using a loop-to-loop method.

have doubled the surface area that the fly line loop is pulling against. It is extremely rare to have a fish run into the backing. In most fishing locations, if a fish gets that far away it will probably be lost in the cover anyway. However, the disadvantage of using the surgeon's loop is that it has a large knot that needs to travel through the guides should you ever have a fish run into the backing.

Recently I have started using the reverse blind splice, a method that creates a loop in your backing without using a knot. Should you have a fish run into the backing, there will be a smooth line running through the guides of the rod. A loop in the end of the backing will take advantage of the welded loops that are on most fly lines today. The fly line can be attached to the backing doing a loop-to-loop connection. It then becomes easy to change fly lines to meet different fishing conditions.

FLY LINES

Modern fly lines are amazing in their diversity and function. Lines are now more specialized than ever. The key to having a line that will deal well with large wind-resistant flies is in the taper of the line, along with the weight of the line. The line must have an exaggerated weight-forward taper. Having the weight-forward taper puts the larger mass at the front of the line, making casting a wind-resistant fly easier.

Rio, Scientific Anglers, Cortland, Orvis, and other companies all produce weight-forward specialty lines designed to cast large wind-resistant flies. The shape of the weight-forward taper as well as the weight of the line will affect how well it will load the rod (loading the rod simply means how much the weight of the line will cause the rod to bend). This in turn will determine how well it will transfer the energy of the moving line through the leader and into casting the large fly.

Rod and line weights can be confusing, especially today with faster graphite rods and matching lines that properly load these rods. Line weights were codified back in the 1960s when the American Fly-Fishing Trade Association (AFFTA) created a standardization of line weights measured in grains. Grain weight was the foundation of the English weight system and was the unit of weight that equaled one grain of wheat. It is equal to 0.065 gram. The AFFTA system measures the grain weight of the first 30 feet of line and

I use a dedicated bag for extra fly lines. Utilizing a reverse blind splice loop in the backing, it takes only a few minutes to change out fly lines for different fishing situations. I recycle all my plastic line spools and instead store lines in these organizers. I cut out the identifying label and keep it with the line, and I store all the labels for the lines in use in my bag and then attach them when the line is changed. The red spool in the lower right corner is a Scientific Anglers Regulator Spool. I use it to change out lines. If you plan on carrying extra lines, this saves on the cost of having to buy extra spools for your reels. It also makes it easier to test different lines to dial in one that works well for a particular rod and fishing situation.

assigns it a number. For example, a 6-weight rod is rated for a line that weighs between 152 and 168 grains, with 160 grains considered the sweet spot. This means that if the first 30 feet of line weighs 160 grains, a rod identified as a 6-weight should cast that line easily.

Today not many rod makers or line manufacturers adhere to the old AFFTA standards for these larger lines and rods. As an example, my Sage Payload 6+ is rated for casting lines that weigh between 200 and 230 grains. I am casting a Scientific Anglers Amplitude Bass Bug line, which weighs in at 210 grains. This line really does hit the sweet spot on this rod. Under the AFFTA standards, both this line and rod should be rated as an 8-weight setup; however, both the line and rod are rated as a 6-weight. In line weights that are considered 6-weight and up, the lines tend to be heavier than the AFFTA standard would suggest. However, with trout rods in the 1-weight to 5-weight range, both rod and line companies tend to follow the AFFTA standards more closely. I don't know why there is this divergence from the AFFTA standards in the larger weights, but it definitely exists.

One of the advantages of having a welded loop at the end of the line, along with a loop in the backing, is the ability to easily change fly lines. Even within a single line-weight designation, there will be some variation of the grain weight from line to line. By easily being able to change fly lines, it is easier to cast a few lines to determine which line seems to best suit the rod you are casting. For example, I also fish with a 7-foot, 10-inch Redington Predator that is rated for an 8-weight line, but I am currently casting a 9-weight line and I like the way the heavier line loads the rod. Most rods that are rated as a 6-weight or above will handle a line that is one weight heavier than the recommended weight for that rod. Experiment with different lines to find one that works for the rod and flies that you are planning to cast.

Cleaning fly lines used for warmwater fly fishing is especially important, as the line tends to get dirty quicker than in coldwater trout fisheries. With some of the new coatings on fly lines, manufacturers are recommending that the line just be washed using a gentle soap diluted in a bucket of water, then rinsed and dried. With other lines it is recommended that a cleaning and polishing product be used. I can tell when a fly line is getting dirty by both the sound it makes going through the guides and how it shoots through them. Given the price of modern fly lines, it makes sense to maintain them.

LOOP-TO-LOOP SINK TIPS

Twenty years ago, I carried numerous extra spools loaded with various sink-tip lines for both my 7-weight rods and 8-weight rods. I would then change spools to match the line I needed for the fishing conditions. This was an expensive way to add sink-tip lines to my fishing tackle, not to mention the additional space needed to accommodate these extra spools.

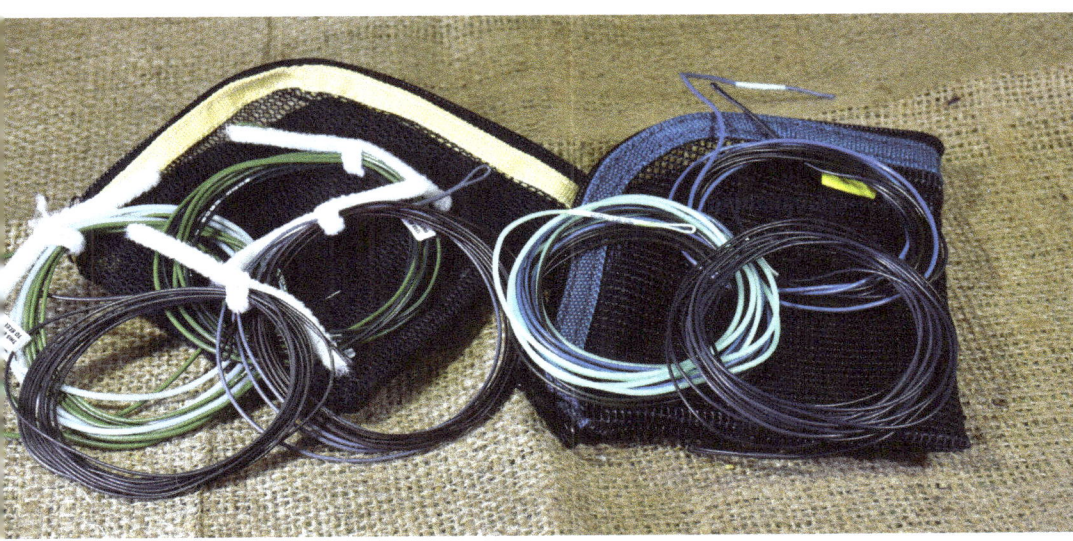

These sink tips were designed for Spey fishing; however, they work equally well when added to a bass taper line. Lines can be quickly changed from floating line to sink-tip line using loop-to-loop sink tips. This is a much more economical option to having multiple spools with dedicated sink-tip lines.

When line companies began to put welded loops at the end of fly lines, I began experimenting with some of the loop-to-loop sink-tip leaders and line tips. All of these various sink tips and "add-on" leaders were designed for Spey fishing; however, I found that they were just as versatile when used with the specialty warmwater lines that I was using to fish for bass and other warmwater fish. The first ones that I used were Rio's VersiLeaders. These leaders have a sink-tip coating in various densities on a monofilament core, and most of them are in the 10- to 12-foot range. They extend my fishing by quickly converting my floating line into a sink-tip line. The only disadvantage is that they are designed to be leaders, so I have to build out the fishing end into a leader; consequently, I cannot stack two VersiLeaders to create a longer sink tip. I still use them, though, as they quickly provide a way to fish a little deeper.

Now I have actual sink-tip sections of line that both Rio and Scientific Anglers manufacture. These sink-tip sections of line, which were originally designed for Spey fishing, have a welded loop at both ends and can quickly convert a floating line into a sink tip. Additionally, I have found that I can add two sink tips to the end of my floating line to create a 20-foot sink-tip line. I use the lighter-density line first then add the heavier-density line to create a line that will maintain a straight-line connection to the fly as it sinks. To the sink tips I will then add a short leader of monofilament or fluorocarbon as a butt section, and then add a section that will be used to attach the fly.

These sink tips have been a real boon to my fishing, as they are relatively inexpensive when compared to purchasing spools, backing, and full-length dedicated sink-tip fly lines. They also take up less space, as I can have a wallet with numerous sink tips sitting in a side pocket of my tackle bag.

LEADERS, TIPPET, AND CONNECTORS

Leaders with matching tippet make the final connection to the fly. Leaders designed for casting larger flies will normally be designated as bass leaders. All leaders are tapered from the butt to the tip of the leader. Bass leaders have a larger-diameter butt section than do most trout leaders of similar size. The tip sections of bass leaders as well as the spools of tippet are designated with the pound test of the material rather than the usual X designation that trout leaders use.

With the X designation system, the diameter of the material is identified. The X designation added to the diameter of the material (in thousandths of an inch) always equals 11. For example, the diameter of a leader that is designated as having a 4X tip will measure out at 0.007 inch. So, 4 + 7 = 11. What can vary is the pound test of the material. Different manufacturers' leaders

Bass leaders have a heavier butt section than trout leaders. The heavier butt is more efficient for casting larger flies and having the leader and tippet land straight.

Instead of using the X designations, bass leaders use pound test to indicate the strength of both the tip of the leader and the tippet being added.

will have slightly different pound-test measurements with the same diameter. With bass leaders measured using the pound test of the tip, the diameter of the tip section will vary depending on the brand.

There are a few considerations that govern the choice of which pound-test leader to use. Larger flies will require heavier tippets. If the leader and tippet choice does not prevent the fly from spinning around during the cast, a larger size needs to be used. Another consideration is the kind of cover that is being fished. If weeds or woody cover is present and the fish needs to be turned quickly to avoid getting tangled, that would also dictate a heavier leader. For bass fishing, a 10-pound-test leader is normally sufficient for most fishing situations. If I am using large bass bugs (e.g., bugs tied on a size 2 TMC 8089 hook or larger), I will jump up to the 12-pound-test leader. For

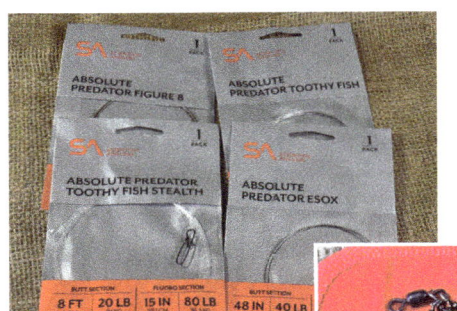

Instead of tying your own leaders for pike and musky, you can purchase commercially made leaders designed for these fish.

Adding a connector to the end of the tippet creates a hinge, giving the fly more action. Warmwater fish are not nearly as tippet-shy as trout.

I use Maxima Chameleon to tie short leaders for sink tips. If fishing for bass, a fluorocarbon spinning line can be used, and if fishing for pike or musky, wire bite tippet can be added. These leaders will have only three sections and will be about 6 feet long.

This bag contains an assortment of Maxima Chameleon monofilament in various sizes and hard saltwater fluorocarbon used for bite tippets, as well as different sizes of wire for the same purpose. I also have a variety of VersiLeaders, as well as extra leaders and tippet. I always carry this bag in the boat so I can quickly create a custom leader for different situations.

panfish the 8-pound-test bass leader is a good choice; however, a 7½-foot or 9-foot leader with a 4X tippet will also work well. I rarely will use a tippet that is finer than 4X for any warmwater fishing situation.

Another type of tippet that is needed for both pike and musky is some form of bite tippet. I have used wire as well as hard saltwater fluorocarbon for this purpose. Both 20-pound-test and 30-pound-test wire work well. When attaching wire to monofilament, use an Albright knot. With hard saltwater fluorocarbon as a bite tippet, use a slim beauty knot to attach the fluorocarbon. The fluorocarbon bite tippet should be between 30-pound test and 40-pound test (perhaps a larger-pound test will be needed in some situations). Wire is the go-to material for pike and musky, but if the musky are following but not taking the fly (a situation that is all too common for this fish), switch from a wire bite tippet to a fluorocarbon bite tippet.

When creating a leader for floating or diving flies, use a 16-pound-test bass leader and add a bite tippet to the leader using appropriate knots. When creating a leader for a sinking line, start with 25-pound or 30-pound Maxima Chameleon monofilament (or some other brand of monofilament). To that attach a section of 18-pound or 20-pound fluorocarbon line. For the tippet, match the fishing conditions or the type of fish and use an appropriate material. For bass, 10-pound or 12-pound test of either monofilament or fluorocarbon is normally what is needed. For pike or musky, use wire or hard fluorocarbon for the bite tippet. Some companies offer premade leaders with an appropriate bite tippet for pike or musky. I have used leaders from both Rio and Scientific Anglers with good results. These leaders are a good alternative to making leaders for these fish.

One of the innovations that has come to fly fishing, which is sometimes more associated with conventional tackle fishing, is the use of connectors at the end of tippets. These connectors come in a variety of designs including snap connectors, clips, and twist clips. Used in various sizes, these connectors can facilitate quickly changing flies but, more than that, they also improve the movement of the flies in the water. With waking minnow–type patterns, they will enhance the side-to-side or "walk the dog" motion. Diving flies will dive deeper, and bottom flies, which are weighted at the eye of the hook, will sink to the bottom with the head down in a vertical motion and the fly can be fished with a jigging motion. Because the clips provide a free-hinging joint, all flies will have better action in the water. They also work well at the end of bite tippets, since you do not have to redo the knot each time you change flies. I do not use them when fishing a topwater popping-style bass bug or for smaller panfish flies, but for many other fly patterns they really increase the action of the fly.

TACKLE BAGS

I love tackle bags! I am constantly working on my tackle bag to either reorganize the items in the bag, clean it out after a fishing trip, or simply inventory what is inside. Tackle bags are really a must-have item unless you fish with minimal tackle. Bags are a personal choice, and as long as it stores your

A well-organized tackle bag will make a day of fishing more efficient and enjoyable.

tackle in a way that is easily accessible, that is all that is needed. Along with the tackle bag, I also have several Fishpond Headgate tippet holders. Since I carry several types and brands of tippet, it is good to have them organized on tippet holders that can be attached to the outside of the tackle bag for ease of access.

CLOTHING

Most clothing designed to be used for saltwater fishing works well for summer fishing. I especially like the new solar shirts. While the intensity of the sun is not as great as when on a saltwater flat, it is always wise to protect yourself from too much sun exposure. This material is cool in hot weather, dries quickly if it gets wet, and also works well during cold weather under a fleece or hoodie. Normally, I fish in shorts unless I am in the Upper Peninsula of Michigan or some other place where bugs are a problem. Then I go to a pair of lightweight saltwater pants. When fishing during the summer, it is always wise to have a second layer such as a fleece top or sweater to use. Even in the best of weather, early mornings and late evenings can get quite cool.

One of the best innovations to come along in recent years is the Lycra finger guard. I would always get line burns on the inside of my right-hand index finger if I had to strip in line fast or if a fish ran quickly. These finger

No matter where you are fishing, it is always wise to take proper precautions to protect yourself from the sun. In addition to appropriate clothing, I occasionally wear gloves to protect my hands and fingers from both the sun and fly line. Before stripping guards came along, I would spend the summer with small cuts and abrasions on the inside of my index finger. I consider stripping guards to be essential. In addition to protecting your index finger from the line, they will pick up water from the line as you fish, allowing the line to glide smoothly across your finger as you strip.

guards provide excellent protection from getting injured, plus the line will slide smoothly across the Lycra material. If needed, a pair of fishing gloves can also be used.

Hats and sunglasses are also needed. A hat will shade your eyes, and a pair of polarized sunglasses will allow you to see below the water surface in clear lakes or rivers. Additionally, sunglasses are simply a must-have item to protect your eyes from an errant cast that could have the fly hitting your face.

FLY BOXES, TOOLS, AND OTHER ACCESSORIES

If you are fishing for all the species that are included in this book (which does not include all the warmwater species of fish that are available), that will involve having a wide range of fly sizes. Trout-size boxes will accommodate most flies for bluegill and the smaller ones for bass, but for the larger flies for bass, pike, and musky, bigger boxes will be necessary.

Other tools that are needed include a good pair of nippers for trimming leader material. Just as with trout fishing, a pair of hemostats is helpful for unhooking fish, especially panfish. For bass, pike, and musky fishing I also carry a pair of fishing pliers for unhooking flies as well as a set of jaw spreaders and a BogaGrip. I rarely use the jaw spreaders, as a pike or musky will normally be hooked in such a way that the fly can be removed using the fishing

Assorted tools (clockwise from top): a lanyard with line clips, hemostats, and a Leatherman tool with scissors; a hook file; a pair of small tool pliers and a pair of side cutters for working on wire bite tippet; a BogaGrip; a set of jaw spreaders to retrieve a fly taken deeply by a pike or musky (use these only when necessary since they can damage the fish's mouth); a pair of angling pliers used mainly to unhook flies from pike and musky; a thermometer to keep track of water temperatures.

A BogaGrip is useful for weighing fish and can also aid in landing toothy fish. Remember to support the fish with two hands if holding it in a horizontal position.

pliers. However, from a safety standpoint, if either of these fish is deeply hooked, using a set of jaw spreaders is advisable to remove the fly safely. I also have a set of small tool pliers that includes a pair with serrated jaws, as well as a pair of side cutters, for working on wire bite tippets.

Fly floatant is another necessity for keeping deer hair bass bugs, as well as dry flies, floating. I also use a lanyard to keep my tools available when fishing from a boat. Attached to the lanyard is a pair of nippers, hemostats, and a small Leatherman tool that has scissors. Since I do not want to overload the lanyard, these are the only three tools that are attached to it. I know some lanyards come with more items attached, so this is another of those personal preference decisions that needs to be made.

I will freely admit that I am a tackle geek. I think that most fly fishers are. I really enjoy doing maintenance on my reels, lines, and other tackle items. How and what an angler carries as well as how it is organized is unique to each of us.

BOATS

As a trout fisherman, I make good use of my waders. Most of the rivers that can be floated also can be waded. Smallmouth bass also are often in waters that can be waded, and ponds can be fished from the shoreline. However,

Personal pontoon boats provide safe and easy access to smaller lakes and ponds.
PHOTO BY FRANK WHISPELL AT WWW.FFFADVENTURES.COM

most other warmwater fishing opportunities require some type of a boat. Even small lakes that are less than 100 acres are better accessed with a boat.

When considering what type of boat is needed, the type of water that will be fished is the main concern. If fishing mainly ponds, a float tube or small personal pontoon-type boat is all that is needed. When I first started to use float tubes, the name was appropriate since it really was an inner tube inside of a cover. It was difficult to launch, plus it only had the inner tube for flotation. While inflation problems were rare, if there was a problem, a quick exit to shore was needed. The modern float tube is normally a U-shaped tube. These U-boats are much easier to launch, and they are also much safer since normally they have two separate inflation chambers and a foam seat for added flotation. Small personal pontoon boats have the advantage of being able to propel them with oars as well as swim fins, plus a trolling motor can be added for increased range.

I have used both styles of boats, and they work well for the right conditions. The advantages are that they are not too expensive and are easy to transport to fishing locations. Also, since the angler sits lower in the water than in most other boats, fish are not spooked as easily. However, one of the great disadvantages of float tubes is that fishing needs to be done relatively close to the launch site. Paddling even a mile in a float tube boat is a lot of work. If fishing from a pontoon-style personal watercraft with oars, the range of area that can be fished is considerably larger but still limited.

Fishing kayaks are also popular in that they will hold plenty of equipment and normally have a larger range when compared to float tubes or pontoon boats. Some kayaks come with a pedal drive where you can use your feet to pedal a paddle system at the rear of the boat. Kayaks are similar in price to float tubes and pontoon boats and offer another option for getting out on the water to fish.

For most fishing situations, some type of boat is needed. Michigan, as well as the surrounding states, has an abundance of lakes that are 500 or fewer acres. A boat is needed for these lakes. There are a lot of options when it comes to deciding on a boat, and cost is a major consideration. Unless fishing in very large reservoirs, a big, expensive bass boat is normally not needed. Flat skiffs are becoming more common and are generally more suited to fly fishing than traditional bass boats. Simple 12- to 16-foot fishing boats or johnboats are probably a good choice for most lakes.

Boats become very personal fishing tools. One of my boats is a 14-foot aluminum boat that I inherited from my grandfather, and it works well on all the lakes I fish regularly. I have a 1960 5.5-horse Evinrude outboard with a bow-mounted electric trolling motor. It's probably worth less than one of my fly rods but has served me well for several decades on the small to medium lakes that I most often fish.

Considerations for Boats

There are a few considerations that make a boat a more effective fishing platform. **The first consideration in boating, however, is always safety!** Always carry Coast Guard–approved life vests as well as throw cushions. Wear the life vest if you think your ability to get to shore, should something happen, is at all in question.

Also keep an eye on the weather. If a thunderstorm is in the vicinity, you should head to shore immediately. My rule is that if I hear thunder, I will leave the lake. I will fish through all-day light rains, but lightning is a different matter altogether. With smartphones available today, combined with apps that give access to weather radar, keeping track of approaching weather fronts and storms is easy and advisable.

Standing up to cast is helpful although not required when fly fishing. While you can cast from a seated position, fly casting from a standing position is far more effective. Having a flat, stable platform or deck to stand on makes casting easier. Additionally, casting from a standing position makes sidearm casts possible. I often use sidearm casts to circumvent wind or to skip a fly under a dock or overhanging tree branches. Bass boats, flats skiffs, and johnboats all provide flat decks to stand on. For the DIY angler, a flat deck can be made and installed on smaller fishing boats. Be mindful of safety and install the deck low enough in the boat so that it is a stable platform to stand on.

A quiet boat is essential to fishing success. If fishing out of an aluminum boat that has a bare deck, install marine-grade or outdoor carpeting on the

Carpeting will help to muffle movements in a boat, especially aluminum boats.

Loose rods can easily be damaged. If your boat does not have rod storage built in, consider adding something to protect your fly rods. I have added a simple PVC pipe to protect the rods. Also note that the rods are in Scientific Anglers' braided polyethylene sleeves to protect their guides and finish. I also always keep the neoprene reel covers on the rods until I am ready to fish them.

There are advantages to standing up to cast. However, a flat deck and stable boat are needed to stand.

deck and gunnels where you stand to fish. This will muffle the noise that can be transmitted to the water.

Consideration should be given to storage of fly rods. Many boats will have rod storage already built into the design. Again, for the DIY angler, homemade rod racks can be built into a boat. If no rod racks are available, simply be mindful of where you store the rods; breaking a rod is frustrating, not to mention expensive.

Having a bow-mounted trolling motor allows you to move along shorelines to fish. Avoid using a motor that has a foot controller, as the controller will catch the fly line. An uncluttered flat deck is preferable for managing fly line. While not a necessity, having an outboard motor certainly

All stern-mounted electric trolling motors have the capacity to reverse the head to facilitate mounting the motor on the bow. Small lakes do not require large outboards. This is a 1960 5.5-horse Evinrude outboard that is sufficient for most small lakes.

A small anchor is normally all that is needed to remain in a stable position to fish. Two anchors are needed if you want to position the boat so it cannot be affected by the wind.

A droguing chain allows you to control the drift of the boat during windy conditions.

shortens your travel time to and from fishing sites. Small outboard motors work for most small to medium lakes. Always consider the travel time back to your takeout point and be conscious of weather conditions. The advantage of a larger outboard is the ability to escape threatening weather as well as allowing you to access more quickly the larger lakes or longer river systems.

Having an anchor can be useful in windy conditions. An anchor allows you to position the boat on a particular fishing spot. Having two anchors is sometimes useful in positioning the boat in a certain way to make it easier to cast in a specific direction.

The most useful tool to control the drift of a boat in windy conditions is a droguing chain. "Drogue" is derived from the Middle English word *dragge*. As the name implies, a droguing chain is dragged behind the boat. There will be more days than not when wind will be an issue. If the wind just pushes your boat along, you will not be able to effectively fish an area. By dropping a chain attached to the boat with a piece of anchor rope, the boat can be slowed down. Having several lengths of chain to customize the weight that is being dragged is needed to adapt to varying wind conditions. Using the trolling motor along with the right amount of chain, the drift of the boat can be controlled so that when the motor is turned off, the boat will come to a stop. This is the method I use most often, although sometimes I will use just enough chain to allow the wind to move the boat slowly, fishing as I drift along.

The one caution in using a chain is to consider the bottom. I have occasionally had the chain hang up on a stump or some other solid piece of structure. In a lake this is normally not a dangerous situation, as the chain can be unhooked by reversing course and moving past the point where it is caught. I would be incredibly careful if choosing to use a chain in a river system, as you have the added hazard of current. I would personally not use a chain with my regular fishing boat in a river, since the boat is not designed for river fishing. ■

4

Fishing Warmwater Flies

Until man is redeemed, he will always take the fly rod too far back.

—Norman Maclean, *A River Runs Through It*

One of my first experiences of catching a bass with a fly rod happened from our dock on Lake George. Using a 7½-foot, 4-weight rod, I was mainly practicing my casting with the bonus of hooking and playing the numerous panfish that hung out in the shallow water around the dock. As the fly was being retrieved, I hooked a larger fish out in the deeper water. It immediately felt bigger than the small panfish that I had caught up to that point. Since I was only a beginning fly fisher, up until then my largest trout had been in the 12- to 14-inch range. This was the strongest fish I had hooked to date.

Reeling up my extra line, I played the fish from the reel. When it got near the dock it suddenly took off for deep water, running about 15 yards of line off the reel. That had never happened before. After a few more runs, I finally landed a 14-inch largemouth bass. On the small trout rod I had been using, it was a thrill. This was the event that convinced me I needed to start paying attention to what was literally out my door as opposed to driving the hour or so to fish for trout.

A largemouth bass caught on a deer hair popper.

47

APPROACH

One of the most influential magazine articles of my early days learning to fly fish was Ken Miyata's "Fishing Like a Predator." Like many of Ken's articles, the title really says all that is needed to understand the concept. Watching blue herons, the fishing method they most often employ is simply standing still and waiting, then quickly striking and capturing a fish as it swims past. This lesson should be well learned by good anglers.

I have sometimes observed some of the local competitive bass fishermen go charging into an area with their bass boat at full throttle. Turning off the outboard, they then jump into the front of the boat, as it settles into the water coming off plane, and begin to frantically cast at the shoreline. The one universal trait of this tactic is they never catch a fish on those first casts. Izaak Walton's centuries-old and final piece of advice, "Study to be Quiet," included at the end of *The Compleat Angler*, would be well learned by these and all anglers.

Approach all fishing situations slowly and quietly. Whether you are wading or fishing from a boat, this advice applies. I relearned this lesson while on the Florida Bay flats fishing for redfish. I had mistakenly worn a pair of low-cut wading shoes with hard bottoms. Several times when I would take a step in the boat it created enough noise to alert the school of redfish and they would change direction. Now I only wear foam-bottom shoes, which are virtually silent when moving about in a boat. I have also carpeted my personal boat

Great blue herons are a common sight around the Great Lakes. This bird instinctually knows how to "fish like a predator."

to make sure any noise from movement inside the boat is minimal (see the chapter 3 sidebar "Considerations for Boats").

When wading or even walking the shoreline, move slowly and observe the water as you go. Often when I step into a river, I will simply stand still for a few moments to observe how my entering the water may have affected the fish. This also allows me to see where the fish are as well as their activity.

When operating a boat, do not use the outboard motor to drive right up to a fishing position. Instead, cut the outboard out in the lake and use an electric trolling motor or row (or paddle) to get into fishing position. When I am working along a bank and know that I am approaching a particular good piece of cover, I will even stop the trolling motor and allow the boat to glide into position.

In addition to fishing so as not to be heard, fish so as not to be seen. Clothing is also a consideration in approaching fish. Wear colors that will blend into the surroundings. That doesn't necessarily mean camouflage but simply colors that will not stand out. When fishing in rivers, I often wear a camouflage shirt and hat; however, out on a lake, shirts that might blend into the sky colors can be worn. Never wear brightly colored hats or shirts.

Similarly, when possible, always fish from a shaded spot. This holds true whether fishing from a boat or wading. In early morning, I always try to start on the eastern shorelines when fishing lakes. Most lakes have large trees lining the bank and the shade line can be quite extensive coming out from shore. Fishing from the shaded side of the lake improves your chances twofold: first you are harder to see, and second the fish tend to be more aggressive. This same principle holds true for the western shorelines in the evening.

Equally important is to limit your false casting. With the large rods used for bass, pike, and musky, the lines are designed to pick up and, with one or two false casts, shoot the line to your target. Also, do not rip a bad cast off the water. If the fly lands somewhere other than where you wanted it to, fish out that cast before casting again.

The one universal principle here is the quieter your approach and less conspicuous your presence, the better your chances are of catching fish.

CASTING

The first fly rod I owned was a telescoping 8-weight with a level 8-weight line. I got this setup at our local Kmart store. At the time all I knew about fly fishing was that you had this rod and this heavy line, which was very different from the spinning and bait-casting rods I had used as a kid, and you used it with this piece of monofilament attached to the end to cast the flies. My fly selection, also from Kmart, consisted of a gaudy collection of some hot pink flies and chartreuse flies. I may have had some more-discrete colors, but the bright ones are the ones I remember. At this point even the concept of dry

Using a double haul to cast flies for musky and pike on an undeveloped lake in the Upper Peninsula of Michigan.

flies or wet flies was foreign to me. However, the one thing I did realize was that getting the fly out to the fish was going to take a very different skill set than what I possessed at the time.

I set the rod up and took it to the backyard and began practice casting. Other than knowing that I needed to get this line to straighten out to put the fly out to the fish, I had no real idea of what casting a fly rod even looked like. Of course, this was before the internet, VHS tapes (or DVDs), and other forms of readily available information. Fly fishing, at the time, was an obscure

sport. For a while my practice session mainly involved waving the rod around and having the line scroll in various S-shaped configurations, with the line piling up whenever I laid it down.

Suddenly, quite by accident, I waited long enough for the line to straighten out behind me. When I came forward with my front cast, I felt the rod bend, stopping the rod on the forward cast, and the line shot out forward and completely laid straight out on the lawn. The feeling of that cast is still etched into my muscle memory almost five decades later. It was an absolute moment of enlightenment of what fly casting should be.

I have been very fortunate over the years to be in the right place at the right time. I have taken classes and worked with some of the finest fly casters in the country, and to this day I am still trying to refine my casting stroke. I have a deep and abiding affection for casting. If fly fishing consisted only of casting, I would still probably cast a fly rod, and the fact that it is a means to an end only makes it better.

Although related, the skill set needed to successfully cast and catch trout on a river and cast larger bass bugs or flies for bass, pike, or musky are different. Casting and fishing flies for trout, especially dry flies, requires precision casting, placing both the fly and line where needed to get a drag-free float. These casts rarely exceed more than 30 feet; however, the fly must sometimes be placed in a trout feeding lane that is only a few inches wide. Except for panfish, most warmwater fishing requires the use of larger line-weight rods and casts that are at least 50 to 60 feet long. Although precision is needed here as well, since placing a fly precisely along a piece of cover is often the difference between catching and not catching a fish, the skill set needed is different.

I am not going to delve too deeply into casting here. As humans we learn in a variety of ways. First there is audiovisual learning, what happens in books, videos, and watching a casting instructor demonstrate a cast. This is an important component, since it imparts the knowledge of what is needed to complete the task. Then there is kinesthetic learning, which is learning by doing. Anytime a physical skill is being learned, it is developed by repeated and correct repetition of that skill. Whenever I am teaching a casting clinic, one of the most important aspects of that clinic is the kinesthetic portion of instruction. Always with permission, I will cast the rod with the student so they can get that feel of the rod bending (or loading), then stopping the rod and feeling it cast the line. Just as it was for me so long ago in my backyard, that sense of loading the rod and then stopping it to let it cast the line is always the breakthrough moment for a beginning fly caster.

There are entire books written about casting as well as many videos, both DVDs and on the internet. I have listed many of them in the bibliography. All casting instructors have their own way of describing how to cast a fly rod. The terminology may be different, but the basic principles of casting are the same. While not trying to do a complete dissertation on casting, I am going to delineate some of the fundamentals of casting as well as describe some of skills that are needed to fish warmwater flies.

First, use what is called a "shake hands" grip. This entails keeping the thumb on the top of the rod handle and the other four fingers curled around the bottom the grip. When using rods that are 7-weight to 11-weight, having the thumb on the top of the rod will make it easier to cast, since the thumb is the strongest digit on the hand. Also, by having the other fingers on the bottom of the grip, the forefinger (and perhaps the middle finger) can be used to control the line as the fly is being retrieved or when playing a fish.

Note the angle the reel-seat portion of the rod forms with the lower forearm. This angle should remain consistent throughout the cast. In other words, keep a very firm wrist, not allowing it to flex, throughout the casting stroke. Most of the problems in a casting stroke come from using the wrist to cast the line.

Make certain to pause and allow the line to straighten out on both the forward cast and the backcast. If the cast is made before the line is straight, the energy of the cast will be dissipated in unrolling the line rather than having the line move in the direction of the cast.

A beginning fly caster should mainly work on the basic false-casting stroke. Everything else you want to accomplish with a fly rod proceeds from having a correct basic casting stroke. Also, if you are an experienced fly caster, continue working on your basic casting stroke as well. Even experienced casters will have difficulty coming to a sudden and complete stop on the backcast. Most casters will do a small drift to a stop instead. An efficient stop will put the most energy into the backcast. Having a correct, high-energy backcast allows for an efficient forward cast.

Practice shooting line (this means extending the length of the line that is being cast). The goal should be to shoot the maximum amount of line with each false cast. This will allow the fly to be delivered to the desired location with the least amount of false casting.

When picking up the line and changing the location that you are fishing, use the backcast to change direction, not the front cast. In other words, sweep the rod around either way to have the backcast travel 180 degrees from the direction you want the forward cast to travel. This eliminates multiple false casts to change direction and will deliver the fly with a straight line faster to the desired location. Part of casting practice should include changing the direction of the cast along with shooting line. Use multiple targets to improve your casting accuracy as well. The quicker and more efficiently you can change distance and direction, the more fish you are going to catch.

Learn to use a double haul! Most of the rods that are used for bass, pike, or musky are heavy and use lines that have extreme weight-forward tapers. Most casters will not be able to properly load (or have the rod bend) with just a normal casting stroke. Using a double haul will put a deeper bend into the rod, which will in turn cast the line with greater speed. Simply put, a double haul cast means that as the rod is loading (or bending) against the weight of the line, you simultaneously use the hand holding the line to pull the line against that pressure of the bending rod. This in turn will force the rod into a deeper bend, and that added energy will then propel the line forward with greater speed. However, describing and doing are different things. You will need to either observe someone, through a video or in person, demonstrating this cast or even better become part of a casting class where the double haul is being taught. One of the idiosyncratic characteristics of using a double haul cast is that it will exacerbate whatever flaws you have in your basic casting stroke. Dialing in your basic casting stroke is a necessary precursor to learning the double haul.

Finally, attend a casting clinic and get good casting instruction. Casting is such a dynamic endeavor that it is difficult to master by just watching videos or reading about it.

Then practice! If, like me, you enjoy casting, practice should not be too onerous. Rarely does anyone improve his or her casting by fishing. When fishing, the focus is on fly selection, where the fish are, and trying to present the fly so as to catch a fish. Rarely is it on the subtleties of the casting stroke. If you really want to improve your casting, you need to remove fishing from the equation.

When practicing, be sure you are performing the cast correctly. While the arm really does not have any memory of its own, muscle memory is often talked about when referring to any physical skill. What that really means is the skill can be performed without having to consciously think about it. If you can false cast and carry on a conversation at the same time without the casting stroke changing, you probably have well-ingrained muscle memory for that cast. Be certain that the practice session is ingraining a correct casting stroke into your muscle memory, since incorrect casting can be learned just as easily as correct casting.

Practicing casting does not mean that you are training for an Olympic sport. If you are a beginner and attend a casting clinic, perhaps a total of 10 hours, spread over a couple of weeks, will be enough to be able to go fishing and enjoy the outing. The ability to cast a fly rod is the main hurdle to really enjoying the sport. Of course, if you are like me and enjoy casting, working on refining your casting stroke is something that you will always continue to do.

RETRIEVING THE FLY

In nature rarely does a creature move with a mechanical, repetitive motion in a straight line. So too should the retrieve of the fly not have a mechanical, repetitive motion. There are two situations to consider when it comes to retrieving (or fishing) the fly: are you fishing a topwater or diving fly, or are you fishing an underwater fly?

If fishing a topwater deer hair bass bug, the first thing to do immediately, as the fly lands on the water, is to drop the rod tip to the water level, then tighten the line up to the fly without moving it. The line should be over the index finger of your rod hand when retrieving (or stripping) line in. Never reach in front of your index finger (the index finger of your rod hand) to grab the line. If you do, you go from having two points of control on the line (your hand holding the line and the point where the line travels over the index finger of the rod hand) to having only one.

Normally at this point I simply pause and wait. When the fly hits the water, if a bass (or pike) is close, it will often trigger an immediate response and the fish will take the fly. This can happen so quickly that you may not be sure the fish did not see the fly above the water as it was landing. On the other hand, if the bass was spooked by the fly landing and darts away, it will often return to investigate the fly. If the fish is a short distance away from where the fly lands, often the sound will attract it to investigate what just landed on the water.

If the fly is not taken, give it the slightest of twitches. It's as if whatever this creature is that has just landed in the water is stunned and is beginning to come back to life. As the fish will often come back to investigate what just landed on the water, this slight twitch can trigger an immediate response and the fly will be taken. When fishing for bass, don't always assume that the fly will be taken in a big, violent strike. I have seen very large bass suck

A nice-size bass for northern Michigan that was caught on a quietly retrieved topwater fly.

a bass bug off the surface as quietly as a trout rising to take a mayfly off the surface of a river.

When beginning to retrieve the topwater fly, continue to keep the rod tip at the surface of the water. In this way as you strip line in, the movement will be exactly translated to the fly. Having a tight line to the fly will also aid in hooking the fish when it takes the fly. Make the fly pop so it makes some sound, then once again stop and let the fly sit. If it is still not taken, continue to retrieve the fly in such a way as to make it pop with intermittent pauses. If a slow retrieve does not work, continue to speed up the retrieve but always have some pauses regardless of the speed. Rarely have I retrieved a fly with a continuous and fast motion and had a fish hit it. It has happened, but slower, more intermittent retrieves seem to be a better tactic.

If you are fishing a dry fly or some type of aquatic adult insect such as a dragonfly, simply twitch the fly rather than retrieving it like a bass bug. This is more a series of twitches and pauses. Then pick the fly up and cast it again to a likely piece of cover or a place where you think a fish might be waiting.

When fishing underwater flies, use the same technique of lowering the rod tip to the water when the presentation is made and tighten slightly on the line. However, then allow the fly to sink. I always count as the fly is sinking. I will start with perhaps a count of 10 and then begin to retrieve the fly. If I catch a fish, I know that a 10 count is the depth the fish are at; if I do not catch anything after a few casts, I will count down to 15. I like multiples of

5, but if you like multiples of 3 or 6 or whatever appeals to you, the point is to progressively fish the fly deeper. With a floating line and leader that is about 12 feet long (including both the leader and tippet), water depths of up to about 6 feet can be fished. If deeper water needs to be fished, add a sink tip to the end of the floating line. If, as you count down, you reach a point where the fly begins to hook vegetation on the bottom (or rocks, etc.), reduce the count so the fly fishes just above the bottom.

Fishing underwater flies normally means fishing with baitfish patterns. Much like topwater fishing, start with slower speeds with occasional stops. However, baitfish patterns can be retrieved with a fast motion since this will imitate a fleeing baitfish. Movement like this will often trigger strikes with bass, pike, walleye, and musky.

If bass are not in an aggressive mood to attack a topwater fly or a quickly retrieved baitfish pattern, a tactic I have found to be very effective is to fish a Woolly Bugger very slowly. Just do very small strips of the line and retrieve the fly as slowly as you can without it snagging the bottom. Bass especially find this tactic hard to resist even when they are not in an aggressive mood to eat.

STRIKING AND PLAYING FISH

When I first began to fish for bass, it was not at all unusual that when I had a strike, I would set the hook only to have the bass jump and I would lose the fish as the fly came sailing back at me. Dave Whitlock's *Fly Fishing for Bass Handbook* finally unlocked the mystery. I was trying to set the hook the same way I set it while trout fishing, by simply raising the rod tip. While this technique works well for trout fishing, where small flies and light tippets are being used, it was destined to fail when if came to hooking a bass.

For the past two decades I have been a volunteer at the Michigan Trout Unlimited Youth Camp. While serving as an overall instructor as well as supervising our campers, my primary role has been to teach casting. One of the insights that I had early on with our campers was how little they understood about how much pressure to apply when setting a hook and playing a fish. Often when we would be out on the river and they would hook a trout, they would try to play the fish with the slightest of bends in the tip of the rod.

When holding a 9-foot rod and feeling it bend against the weight of a fish, it can feel like you are exerting a great deal of pressure. However, you are feeling the force from the wrong end of the lever. If someone holds the line and another puts tension on the line by bending the rod, it's amazing how little pressure is being applied at end of the line. Normally the line can be easily held with just the thumb and forefinger with even a deeply bent rod. One of the activities we adapted was to have campers pair up and each take a

Use the midsection and butt of the rod to both set the hook and play the fish.

Most largemouth bass will jump when hooked.

turn holding the rod and holding the line. Once we started to do this activity, our students began to put more pressure on a fish when it was hooked.

Just like our campers, I had failed to understand how little pressure I was applying to the bass that I was hooking. When a bass, or many other predatory fish, eats some form of prey, it first clamps down on it inside its mouth. This is an attempt to both secure the prey and crush any movement. If that prey happens to be our fly, in order to hook the fish the strike must overcome this pressure so as to move the fly inside the fish's mouth. Moving the fly will hook the fish. By striking the fish with only the tip of the rod, the fish was probably never really hooked, so as soon as it jumps and opens its mouth, the fly will come out.

To hook bass and other large predatory fish, you first must be fishing with a low rod tip, preferably at water level. When you see a take or feel the fly being taken, your first movement must be to pull the line sharply back with the line hand. This will tighten the line against the weight of the fish. Simultaneously lift the rod horizontally so that the power of the strike (or bend in the rod) comes from the butt section and middle section of the rod, rather than the tip. This method applies much greater force to setting the hook. Once I had mastered this technique, I rarely lost a bass or other large fish. This technique should be practiced, especially if you do a lot of trout fishing, because like casting, striking a fish properly needs to be more of a muscle memory reaction than something that is consciously thought about.

Once a bass is hooked, one of the first actions it will normally take is to head to deeper water and attempt to jump in order to dislodge the fly. As quickly as you can, try to get the excess line reeled up so the fish can be played from the reel. Use the small finger of your rod hand to control the line as it is reeled up. While doing this you also need to give most of your attention to what the fish is doing and be certain to keep proper tension on the line by stripping it in when needed or letting line go out if needed. Controlling the line becomes much easier once it is on the reel.

One of the principles of playing a fish is to use the rod to apply pressure in the opposite direction the fish is moving. Rarely do you simply hold the rod at an upward angle and let the fish move about. By applying pressure against the direction the fish is moving, it will tire more quickly. With bass this is particularly important when the fish is attempting to jump. When a bass jumps, that is the moment it is most likely to come unhooked. Drop the rod tip down, even into the water, so as to apply downward pressure against the jump. This will greatly enhance your chances of landing the fish. Once the line is on the reel, the drag can also be adjusted to a higher setting if needed.

LANDING A FISH

Once the fish is brought to the boat or if wading, next to you, it is time to land it. Different fish require different considerations when you are landing them.

Most panfish can be lifted from the water using the rod and line. Once the fish is high enough that you can reach it, grab it from the belly. All panfish have sharp spines on the dorsal fin that hurt if they stick your hand. Often

Bass can be landed by grasping the lower lip; however, do not try this method on other fish.

Support the body of any fish that you lift out of the water.

you will also need a good pair of forceps to get the fly out, as most panfish have small mouths and tend to suck the fly in.

When landing a smallmouth or largemouth bass, whether in a boat or wading, slide the bass toward you and solidly grasp the lower lip by putting your thumb inside the fish's mouth. Bass have teeth but they feel more like sandpaper and will not hurt your thumb. I will just leave the fish hanging vertically in the water, unhook the fly, then release the fish. This does the least harm to the fish. If you plan to take a picture, lift the fish vertically from the water. Then if you want to hold it in a horizontal position, support the fish with your other hand before doing so. *Do not* force the fish into a horizontal position by leveraging the jaw only. I know this is often seen on shows, but it can injure the jaw of the fish. Plus, have enough respect for this fish that you just caught to handle it in a way that will do the least damage.

With either carp or walleye, perhaps the best way to land the fish is to grasp the tail, then set the rod down in the boat, or if wading put the rod under your arm. Once your second hand is freed up, you can grasp the belly of the fish and lift it from the water. You should not put your fingers into the mouth of either of these fish. A carp's mouth is very strong, and it will bite down on your fingers; however, they don't have any sizable teeth. Walleye do have teeth, so you do not want to put your fingers in their mouth. Once the fish is secured in the boat or on shore if wading, use either a pair of forceps or pliers to remove the fly.

The fact that bite tippet is needed for pike and musky should be a clear indicator that you do not want to get your hand anywhere near either of these fishes' mouths. There are a couple of methods to land them. Like with carp and walleye, grasping the tail with one hand and then using the other hand to hold the belly while lifting the fish from the water works. Another method is to use a BogaGrip to grasp the lower jaw of the fish and lift it vertically from the water. Just as with bass, if this method is used be sure to support the fish with your other hand to hold it in a vertical position. Removing the fly can also be an issue with either of these fish. Use a pair of angling pliers to remove the fly, and if it is hooked very deeply, use a jaw spreader to hold the fish's mouth open. You really do need the proper amount of respect for the teeth in either of these fish. Finally, don't ever put your fingers inside the gill plate to lift these fish or any fish. The gills are delicate and can be easily damaged.

Of course, you could also just have the appropriate-size net to land all of these fish as well.

RELEASING FISH

Being primarily a catch-and-release fisherman, one of the practices I religiously employ is to bend down the barbs on all my flies. Doing so facilitates the release of the fish as well as occasionally myself.

One summer in late June, there was a guy fishing off the point where our cottage is located. He was close enough to shore that I could easily talk to him. Upon inquiring about the fishing, he began to complain that he was not catching any bass from our point like he had earlier in the month. I asked him how many fish he had caught here, and he replied that he thought it was

In the same way you would release a trout, hold a bass in the water and allow it to swim out of your hand.

about a dozen bass. When I asked him how many he had released, he replied that he kept them all. Bass being somewhat territorial, I then asked him how many bass he expected this point to support. It was as though the thought had not occurred to him that the number of fish in a location was finite. While our lake is a productive one, especially given the amount of development around it, it still has a limited number of bass in any given location and in the lake as a whole. Warmwater fisheries can sustain some fish being harvested, but catch-and-release is an important practice if a quality fishery is going to be sustained. As Lee Wulff said, "Game fish are too valuable to be caught only once."

Casting Tournament-Legal Bass Fly Rods

Under most bass tournament rules, rods must be less than or equal to 8 feet in length. Rod manufacturers began to make fly rods to qualify under these rules perhaps a decade ago. Even if you do not plan to fish in a bass tournament, these rods offer some advantages. Primarily, because of their shorter length, they give the angler greater leverage in fighting a fish. However, if you are used to casting a 9-foot rod, it will take some time to adjust to the shorter length. Here are some keys to successfully casting these rods:

Pick up and cast a shorter amount of line. This is particularly true if casting a line with a heavy head. If the line is marked or colored differently between the head and the running line portion, be sure to get the head of

the line through the tip top of the rod before beginning the cast. If you are hitting yourself with the fly, you are attempting to pick up too long a line.

Limit your false casting. When you pick up the line to cast and change direction, limit yourself to two or three false casts.

Minimize shooting line during your cast to lengthen it. Instead, rely on the weight of the head to shoot line to the target on the final cast. Gravity will take over when trying to cast a long line that has a heavy head, and the line will drop as it is traveling forward. With the shorter rod, the margin of error between successfully casting the fly and having it hit you in the back is small. Use the shooting capabilities of the line to present the fly rather than trying to hold a lot of line in the air.

Double haul! Using a double haul in your cast takes advantage of the heavier head and the power in the rod. There is a lot of power in the midsection of these rods, and by incorporating a double haul into your cast you take

advantage of that power and dramatically increase your line speed. While gravity still pulls the line down at the same rate, if the line is traveling faster, it has less time to fall and less chance of the fly hitting you.

These rods are not made for long-distance casting. Casts from 35 to 60 feet are normally all that is necessary.

I have grown to appreciate these shorter bass rods. They can be very accurate and handle large wind-resistant deer hair flies with ease. Additionally, they are exceptional fish-fighting tools that can turn a very large bass from cover quickly. If you enjoy and do a lot of bass fishing, these rods are worth your consideration. ■

Using a double-haul when casting the shorter bass rods generates the line speed needed to keep the heavier lines elevated and heading toward their target.

5

Largemouth Bass

As a group, bass are arguably the most important game fish in North America.

—Keith A. Jones, PhD, *Knowing Bass: The Scientific Approach to Catching More Fish*

As the summer was waning and early fall was starting to be felt in the air, the time was approaching to begin the process of closing our cottage on Lake George for the winter and returning to Colorado. This is always a bittersweet time for me, as I am sad that, along with summer, my warmwater fly fishing is coming to an end. However, I am, at the same time, ready to return to the mountains and jump into the fall trout fishing.

Traditionally I have always put my boat up on the trailer for my last few fishing days and visit a few of the many lakes that are in the area. Lily Lake, which is only a few miles away, has traditionally been my Michigan fishing finale. As the name implies, the lake is loaded with large beds of lily pads that provide ample targets for my bass bugs.

On this last occasion, I was fishing along the edge of one of these lily pad beds, working my way back to the boat ramp. I came up to a favorite and reliable piece of cover. The water was only a few feet deep; an old, weathered stump rose out of the water with reeds

Lily Lake is aptly named, with the main cover in the lake being extensive lily pad beds along the shore.

behind it and lily pads next to it. It was a classic piece of cover, where several edges came together.

When I cast my bass bug into the open slot of water next to the stump, a bass came out from under the lily pads and sucked in my bass bug so quietly that I almost missed the take. Fortunately, when I set the hook, this very large bass headed for the open water of the lake behind me. It was large enough that as it ran toward the middle of the lake, it pulled the bow of my boat out as well. Only the really big ones can do that. When I landed it, the fish proved to be one of the largest bass of the summer, in that 5-to-6-pound range. For northern Michigan, this is considered a pretty big fish. Any time you hook a large bass, it is the culmination of fishing the right fly, into the right place, at the right time.

ABOUT LARGEMOUTH BASS

It seems that most fly fishers will tell you that they prefer fishing for smallmouth bass over largemouth bass when given the choice. In my fishing experience it is rare that largemouth bass and smallmouth bass inhabit the same lake; however, when they do, the largemouth is the dominant species, and they will chase the smallmouth out of the prime habitat. This does not necessarily mean that from an angling perspective the largemouth bass is a superior fish to smallmouth bass. Even though they belong to the same family of sunfish, they are each unique. The largemouth, when hooked, tends to

I have always thought of fly fishing as an artistic endeavor. The real thing along with Jeff Currier's artwork makes for a compelling image.

Largemouth bass do not school but will travel in loose groups when hunting for prey.

have slower but more powerful runs. They almost always will try to jump to disengage the hook. The smallmouth tends to make faster runs with more head shaking. They will almost always jump as well. I enjoy fishing for both species and don't have a preference for either.

Often largemouth bass are the top predator in the lake. If pike are present, they will be the top predator, but they tend to move out into the cooler and deeper water as the lake temperature begins to warm up, leaving the shallow water to the bass. As described in chapter 2, "Understanding Lakes," bass relate to the shallow-water cover, structure, and edges. Bass behavior relating to the time of year will vary depending on where you are in the country. Bass everywhere move through several stages of behavior in the same order, although not at the same time of year. Water temperature is the key determining factor in this progression.

In temperate regions where the four seasons are distinct periods, bass will begin to become more active during the spring warm-up and pre-spawn period. This period occurs when the water temperature has not yet reached the 60-degree mark. Aquatic vegetation is still sparse, as is aquatic insect activity. A largemouth bass's metabolism will begin to get more active as the water warms up but will still be slow. Feeding response will also be slow at this time of year, requiring anglers to fish slow to match this behavior. Swimming flies such Woolly Buggers or streamers on sink-tip lines will be needed. Retrieves should also be fairly slow.

The spawning period for largemouth bass will begin to occur when water temperatures start to climb into the low 60s. Largemouth bass require a sandy or gravel bottom in shallow water to sweep out nests. If these conditions exist in the northwest corner of the lake, this is where the first largemouth will begin to spawn. Not all bass will spawn at the same time. Different locations on a lake will have bass still in the pre-spawn stage. While bass are in that spawning stage they are not as interested in eating, although they will hit flies as a territorial or defense response. You will have to decide about the ethics of fishing for bass that are sitting on a nest. Normally there are other fish just outside the spawning area and, if I am fishing during this time, I cast to the edges of this area and leave the spawning beds alone.

The post-spawning period follows the spawn. The male largemouth bass will stay on the nest to guard the eggs, but the females will move off to spend time recovering. They will not be actively feeding during this time and fishing will be slow. Not all bass will be in the same behavior stage at the same time, so you can still catch fish. It is just that the fishing will be slower during this period. As the weather warms up along with the lake temperatures and the vegetation matures, bass will slowly transition out of this phase and begin actively entering the summer peak period.

The summer peak period is the prime time for bass fishing. The aquatic vegetation is now at its peak growth, and the water temperatures have warmed up into the mid-70- to mid-80-degree range. Largemouth bass are actively feeding at this time. If water temperatures are stable, this is a prime time to fish topwater bass bugs. While some fish will always be off the drop-offs in deeper water, many will remain in shallow water for most of the day if the water temperatures remain in that 75- to 85-degree range. If the water temperatures climb into the high 80s, the bass that are in shallow water in the morning will sometimes move off to deeper water as the day progresses. Sometimes they return to shallow water in the evening, but normally not in the numbers that are present in the morning.

Another factor to consider when locating bass during this time is shoreline activity. Most lakes have at least some development and if people are out and active on the shoreline, that will normally move the bass out to the first drop-off. Personally, I never fish right in front of a residence where people are out sitting on their dock or deck or moving around on shore. For me, it is just a matter of good etiquette to give people their space. Normally there is a lot of shoreline that is quiet or even undeveloped that can be fished.

An additional factor that can affect fishing during this time is having a cold front move through the area. A light and brief rain can trigger bass to feed. Perhaps the rain oxygenates the water, triggering this phenomenon. However, a sustained rain with wind will lower the shallow water temperatures and bass will retreat to the deeper water. This is not a result of the

Fly Fishing in Canals

With the vast number of lakes available it is sometimes difficult to know where to start. One option is to look for lakes with canals. Often canals are added to lakes to provide boat access for additional homes. If canals are maintained for boat access and have sufficient depth, they can sometimes yield surprising rewards.

Canals are often overlooked by anglers but can be tailor-made for fly fishing. At first glance it may seem that there are few fish or only small panfish in canals. However, do not assume that this is always the case. Many times I have approached a canal thinking of using my small fly rods for panfish, only to spook larger bass who went there for the same reason. Larger predatory fish in these situations are very wary and can be spooked easily. A cautious approach is vital.

When entering a canal, assume that there are large fish present and make a cautious approach. Fish the mouth of the canal before entering it. Topwater deer hair bass bugs are an excellent choice, as often canals are shallow. As described elsewhere, use a slow and subtle fishing technique when fishing a bass bug here. Since large bass have moved away from their deep-water retreat, they are very easy to spook. On more than one occasion I have seen sizeable bass, who had been alerted to my presence, heading out to the mouth of the canal and into the larger lake. However, with the proper amount of caution, it is really a thrill to hook and land a larger fish in these confined waters. ■

Approach the opening of a canal with caution. Large predatory fish are very wary when they are too far away from the deep water. When in the narrow confines of a canal, stop fishing and observe for a moment. You will normally get only one cast to good fish here.

Fishing Heavy Cover

I normally don't tie my deer hair flies with weed guards. If the fly is cast accurately and retrieved carefully, it can be fished around cover without getting hung up. Weed guards will keep a fly from hanging in cover (although even with a weed guard, flies can become tangled). A weed guard will also make hooking a fish more difficult.

Instead of using weed guards on my deer hair flies, I will use a STP Frog fly for fishing in the thickest lily pads. This is a lightweight foam fly with a double mono loop weed guard. When I want to fish in cover that I know I will get hung up in, this fly is a better choice than a topwater deer hair pattern. I find the lighter weight allows the fly to be retrieved through, around, and over cover without getting hung up as often as a deer hair fly would. Also, the clamshell front on the fly will give a good pop in the water when it reaches those open holes in the cover.

When fishing a large bed of lily pads or reeds, use the heaviest bass leader and tippet in your tackle bag. The heaviest tip strength available in a bass leader is normally 16-pound test. Usually this is heavy enough to handle most situations.

Knowing the depth of the lily pad bed is helpful as well. If a fish is hooked and then tangles up in the pads (or reeds), most electric trolling motors will be able to move the boat into the bed to retrieve the fish as long as the water is deep enough.

For this type of fishing, I use one of my heavy-line-weight rods. The rod I normally use is the Sage II Largemouth. This is effectively a 10-weight rod. It's not the fly size that dictates the rod choice, since STP Frogs are very light fly patterns, but the fighting qualities of the rod. It is not unusual to have to move up to the fish and simply dead-lift it up into the boat.

Although not one of my favorite ways to fish, heavy cover is a place where bass (and sometimes other predatory fish as well) can be found. Combining the STP Frog with a stout leader and a heavyweight rod, these places can be effectively fished. ■

The lightweight STP Frog combined with the double mono weed guard allows this fly to be fished in heavy cover like a bed of lily pads.

bass being uncomfortable with the temperatures but has to do with the food chain. When the water temperature decreases it slows the aquatic insect activity, so the smaller fish move out of the shallow water due to decreased opportunities to feed and the bass will follow. A simple clue to how bass are behaving is to check out the panfish activity in the shallow water. If a cold front has moved through the area and the panfish are absent from the shallow water, you probably will need to fish deeper for the bass, and the fishing will be slower. If the panfish have returned to the shallow water and are active, the bass will probably be present there as well.

TACKLE FOR LARGEMOUTH BASS

Suitable rods for largemouth fishing can range from a 6-weight all the way up to a 10-weight rod. A 6-weight is an ideal rod for smaller deer hair flies, and it doubles as a rod that can be fished for panfish as well. I use an 8-weight rod most of the time when specifically fishing for largemouth bass. This weight rod, when paired with an appropriate bass taper line, will handle most all the flies that will be used, as well as a variety of fishing conditions. Rod lengths can range from under 8 feet up to 9 feet long. For more information on tackle, see chapter 3, "Modern Tackle for Warmwater Fly Fishing."

TOPWATER BASS BUGS

I have a deep and abiding affection for fishing topwater bass bugs. Just like fishing dry flies for trout, this is a very visual experience and is quite addictive. If fact, just like fishing dry flies, I will always be slow to change methods when it is not producing fish. Simply casting bass bugs and working at hitting targets is really enjoyable.

Bass feed differently than trout. When fishing dry flies for trout, particularly during a hatch of aquatic insects, you are normally working to match the size, shape, and color of the hatching insect. If the fly selection is correct and you get a good drag-free drift (assuming you are fishing in a river), chances are good that a trout will rise and take the fly. Normally if a trout rises and then refuses a fly, my first tactical change would be to try the same fly but one size smaller. However, bass are not selective feeders as much as opportunistic feeders. Bass do see color and certain colors are always good to work with; however, I am not sure that color is the determining factor in whether a bass will take a topwater bass bug.

The late Doug Hannon, "The Bass Professor," caught his biggest fish between the hours of 10:00 a.m. and 3:00 p.m.—good evidence that largemouth bass are primarily visual feeders. While bass do see color, it is rare that I can just switch to a different-color bass bug and suddenly begin catching fish. Instead, it seems that action is more of a key factor.

This box contains medium-size deer hair bass bugs in assorted colors. This is the box I use most often during the summer months when fishing for largemouth bass. Bass bugs of this size are large enough to attract both bass and larger panfish.

If you have spent much time fly fishing for panfish, perhaps you have experienced catching and playing a panfish that is splashing around on the surface, only to have a bass come charging up from the depths to eat the panfish. Largemouth bass will be reluctant to let go of a small bluegill once it has gotten a hold of it. This can be quite a tug on a light fly rod, but eventually the bass lets the poor panfish go. When this happens, I will normally try to release the panfish in shallow water so it can at least have a chance of recovering before encountering another bass. Other than losing a few scales and looking rather confused (can a panfish actually be confused? probably not), they will normally swim off none the worse for the experience.

The disturbance in the surface of the struggling panfish is what triggered the bass to eat it. The same is true of fishing bass bugs: it's the retrieve that triggers the fish more than the color of the fly. Changing how you retrieve the fly will produce results more often than changing to a different-color bass bug.

With that said, however, if bass are reluctant to hit a topwater bass bug, sometimes dropping down to a smaller bass bug will trigger a fish to strike.

Perhaps the smaller fly is less intimidating, so the bass reacts more aggressively. An angler who is fly fishing for bass in shallow water has a distinct advantage over a lure fisherman who is trying to fish the same water. A deer hair bass bug landing in a couple feet of water normally activates the curiosity of a bass as opposed to a lure landing, which if not skillfully cast can spook fish.

Topwater bass bugs come in a variety of materials and actions. I prefer deer hair topwater bass bugs because when a bass grabs a deer hair fly, it feels like something to eat and the fish holds on to it longer, giving the angler more time to set the hook. When I fish balsa or foam poppers, I tend to fish the flies faster than deer hair flies. The take on these flies with harder bodies tends to be faster, with the bass expelling the fly faster. Consequently, faster retrieves with faster strikes will result in more hookups.

Unlike trout flies, there isn't really any codification of bass bug patterns, although a few fly tiers have attempted to name patterns. Dave Whitlock's Fruit Cocktail Bass Bug is one example of a recognizable fly pattern; however, very few bass bugs have the same kind of recognition as that pattern.

These smaller deer hair bass bugs are fished when bass are reluctant to take larger bugs. Often the smaller size will prompt a more aggressive response from fish. As an added bonus, the larger panfish will also take this size fly.

The left side of this box holds assorted large deer hair bass bugs. I use these flies primarily when I am less concerned about catching a lot of fish and working to catch larger bass. The right side of the box contains an assortment of deer hair bass flies with different-style heads. While I normally use standard-style flies, I also enjoy experimenting with different styles of heads.

If you search for bass bugs on the internet, you will find a large variety of color schemes and styles. This is really reflective of the individual tiers. As I work to fill my boxes for a summer of fishing, I instead think of general color schemes for my deer hair bass bugs. With both my topwater deer hair bass bugs and my diving bass bugs I tend to group flies into three general categories by color, as follows:

Bright colors: This group is composed primarily of white, chartreuse, orange, and yellow flies. These flies can certainly have accents of darker colors, but the overall impression is that of brightness.

Dark colors: In this group I would include black, purple, gray, and natural deer hair bass bugs. Again, there could be accents of lighter colors or perhaps a white face, but the overall impression is that of a darker bass bug.

Frog colors: This group consists of color schemes that have light-colored bellies, various colors for the midline, and darker-colored backs. My favorite color scheme is in this group: the white belly, yellow midline, and olive back frog pattern. This is the color bug that I fish most of the time.

Of course, this is only a general color guideline, which I even break myself. Creativity is one of the gratifications of tying your own deer hair bass flies.

My normal fishing technique is to cast the bass bug and then drop the rod tip and just tighten up the fly. I then let it sit until the small wave rings from the fly landing have dissipated. After that I just give the fly a slight twitch. This is often when the bass will hit the fly. The bass are alerted that something is happening when the fly lands on the water and will investigate. I have seen bass come from up to 20 feet away to investigate a bass bug that lands on the water. If you begin the retrieve too soon, the fly may move outside the window of the fish's interest before it has a chance to inspect it.

DRY FLIES
Another class of topwater flies are dry flies. Although these patterns are more conducive to fishing for panfish, bass will also take dry flies. They can be

This box of dragonfly and damselfly patterns is my Richard Pilatzke box. Richard is a friend and noted Colorado fly tier who can produce a dizzying number of dragonflies during a day at one of the fly-fishing shows. He has been very generous over the years in giving me his dragonflies and damselflies. Almost all of these flies were tied by him. They not only will pull up largemouth bass but also are effective for panfish.

Any large western dry-fly patterns like these Irresistibles will work well for bass and panfish. They can also be fished for smallmouth bass on rivers much as you would fish a dry fly for trout.

either imitative flies or attractor patterns. Lakes have an abundance of aquatic insects, and flies that imitate dragonflies, damselflies, mayflies, and caddisflies are all productive at certain times. However, while fishing flies that are small may work for panfish, they are not as productive for bass. Normally I will fish dragonfly patterns for bass, but the other aquatic insects are probably too small to consistently catch bass. Another productive group of insects to imitate are terrestrials, with grasshoppers, crickets, wasps, hornets, and bees being the most prevalent for bass, as these insects tend to be bigger.

Attractor dry flies are an additional group of patterns that will work well for bass. Any of the attractor patterns or standard dry flies that are tied for trout will be effective if tied in larger sizes. In *Favorite Flies and Their Histories*, Mary Orvis Marbury has a chapter on bass flies. They are essentially the standard trout wet flies of the time only tied in large sizes. Large attractor patterns commonly used in western rivers such as the Irresistible, Royal Coachman, and Wulff patterns as well as other dry flies can all prove productive.

Dry flies are fished differently than topwater bass bugs. Cast a dry fly into a likely looking area around different types of shore cover, particularly shaded areas, and then simply twitch them but do not retrieve them. Often, I will probably be catching more panfish than bass and the bass that I do catch tend to be smaller. However, I am normally using a lighter fly rod, on which even a smaller bass can be an interesting experience.

WAKING AND SLIDER FLIES
Waking flies are tied with rounded heads of either deer hair or another material such as foam or balsa. Such patterns make a subtle wake rather than a pop in the water when retrieved, and include mouse patterns and waking minnow patterns.

With a mouse fly pattern, it is best to do a slow, steady retrieve so the fly swims across the water in a manner consistent with a mouse-like swimming motion. This is an effective way to fish for bass late in the evening as the twilight hours approach.

Waking minnow patterns should be fished in a manner that suggests a struggling baitfish at the surface. If a small fish is injured, especially if it is having issues with its swim bladder, it will be struggling at the surface, presenting an easy meal for bass. The swim bladder in a fish is what allows it to control its buoyancy. Simply waking the fly with a floating fly line and leader is effective, but I also will put an intermediate tip on my bass line and a short leader. The intermediate tip will pull the fly under when retrieved, then the fly will return to the surface when the retrieve is paused. You can control how deep the fly dives by how long you pause between retrieves. The longer you pause, the deeper the intermediate tip will sink, which will result in a deeper dive. I also always use a clip on the fly so the fly is on a free hinge, giving it

This box of assorted waking minnow flies has both attractor patterns and imitative flies. These flies can be fished as topwater flies using a regular bass leader. However, the best way to fish a waking minnow pattern is do a loop-to-loop attachment and add an intermediate sink tip to the end of the bass taper line; then, using a relatively short leader, attach the fly with a clip. The fly can then be fished as an injured baitfish trying to dive back underwater. The clip creates a hinge, allowing the fly to have more movement. This is an effective way to fish over deeper water, as bass, pike, and musky are all very conscious of the movements of an injured baitfish on the surface.

more action than if you just tie the fly to the leader. This simulates an injured baitfish that is struggling to swim but can't escape the surface.

Slider flies, while similar to waking flies, are also part of this group of bass fly patterns. One of the oldest and perhaps best recognized is the Sneaky Pete bass fly. With its pointed head of balsa wood, cork, or foam, it slides through the water creating only a subtle disturbance. There are times when a popping fly pattern is just too much action, and these more-subtle flies are more effective. Slider flies can be tied with deer hair heads, creating a fly that moves through the water with only a minor disturbance.

DIVING FLIES

Diving flies and topwater flies are very similar in how they are fished. I find diving flies to be good at pulling out bass from under overhead cover. Because the fly dives, the fish don't just hear it but can also see it as it dives. I use a clip on all my diving flies. By using a clip, the diver then has a free hinge point at the eye of the hook. This can increase the depth of the dive by several inches.

When tying diving bass bugs, I use both a Dalberg Diver–style diving head and the diving head designed by Tim England (I illustrated tying both flies in my first book, *Tying and Fishing Deer Hair Flies*). The England head is a better diving head, as the hair on the bottom of the fly ensures that the fly lands upright when cast. I have to work at leaving some hair under the hook shank when tying a Dalberg Diver because when the hair collar absorbs water, the top of the fly will outweigh the bottom and the fly will start landing upside down. I treat the collar and top of the head with fly floatant and leave the bottom of the fly untreated. That way the hair on the bottom of the fly along with the weight of the hook bend will help the Dalberg Diver land correctly each time it is cast.

If my topwater bass bugs are not catching fish, I will normally go to a diver fly next. Having the fly dive and then return to the surface is a very enticing action for catching a bass.

This is a box containing assorted deer hair divers. I have both Dalberg Diver–style heads and patterns with a Tim England Diver head. Divers can be fished interchangeably with topwater bass bugs. Sometimes when bass are not responding to a topwater fly, I will first try a diver before fishing deeper. Divers can be fished using a clip, as the hinge effect created allows the fly to go deeper than when tied directly to the tippet.

Innovative Deer Hair Flies

When I first started to tie flared deer hair bass bugs, not many other tiers that I knew were working with deer hair. In addition to sparse information on the techniques, it was also difficult to get quality materials that would work for the flies I was attempting to tie.

The fruits of his labor at the vise: Jacques Bordenave with a nice largemouth bass.

Today all of that has changed. A wealth of quality materials is available now that didn't exist 30 years ago. While deer hair itself has not changed, the way in which it is processed has dramatically improved. Whitetail deer provide some of the best hair for bass bugs, which consists of spinning hair (from the flanks of the deer) and belly hair. The underlying color of the flank hair is gray, so when dyed the colors tend to be more muted. Belly hair is white, so when dyed produces vibrant colors. Other materials such as feathers, flash, adhesives, eyes, and so much more are also readily available.

UV resins are perhaps the most revolutionary product used in fly tying today. They have allowed tiers to create flies that would have been difficult, if not impossible, a decade ago. UV products have made their way into all areas of fly tying, including the tying of deer hair bass bugs.

With the internet and social media, it is much easier to connect with fly tiers from literally around the world. One tier that I might never have met in person, since he lives in France, is Jacques Bordenave. He is one of the most gifted and innovative deer hair tiers that I

A box of Jacques's meticulously crafted deer hair flies.

Two of Jacques's Jithairbugs.

know. Not only are his deer hair flies works of art, but he also engineers them to behave in some very specific ways.

My uncle Lynn used to tell me stories of coming up north after work to fish with my grandfather. The drive up from Midland,

A Jithairbug next to two of the original Arbogast Jitterbug lures.

Michigan, would take about an hour, so they would fish the late evening and into the dark. Fishing jitterbugs along the shoreline was a favorite tactic. I have my grandfather's tackle box and it still contains some of the same jitterbugs that were fished during those evenings.

Numerous tiers strive to tie deer hair flies that look like and mimic the behavior of traditional bass plugs. The jitterbug tops the list. I have made a few attempts through the years and while I can tie flies that look like a jitterbug, getting it to behave in the water like a jitterbug is another feat altogether. Jacques has cracked this code. His Jithairbug is not just clever word play, but making use of quality deer hair, masterful tying techniques, and modern UV resins, this fly both looks like and fishes like the jitterbugs that my grandfather and uncle used to fish.

If you are a deer hair tier, hopefully you will find some of the same inspiration for your own tying that Jacques's fly patterns have had for me. ■

SWIMMING FLIES

Swimming flies are a large category of patterns that can be fished just under the surface all the way to flies that swim just above the bottom of the lake. These flies can imitate baitfish, aquatic nymphs (see the chapter 5 sidebar "Nymph Fishing for Bass"), leeches, tadpoles, and other aquatic creatures. They can also include attractor flies such as large wet flies or traditional streamers. This is probably the most productive category of flies simply because of the diversity of patterns, plus the fact that you are fishing the water column between the two most distinctive edges: the surface of the water and the bottom of the lake. Also, bass, like most fish, do most of their feeding under the water rather than on the surface.

Traditional feather-wing streamers and bucktails are effective for bass fishing, but baitfish imitations with the new synthetic fibers are a more modern look for these flies. With the broad range of colors available, it's easy to imitate the small baitfish in the waters that you are fishing. For bass fishing using colors imitating the small bluegill is very effective. Juvenile bluegill and other panfish are the prime forage for most bass, and these small baitfish are much more subtle in their coloration than the adult fish. Using the range of colors of these new synthetic materials, a realistic imitation can be tied for all baitfish.

The swimming flies in this box are baitfish imitations. Since baitfish are the primary forage for largemouth bass, walleye, pike, and musky, this box can be used for all these fish. These flies will normally be fished using a sink tip and short leader added to the end of the fly line.

The Woolly Bugger is a go-to fly for bass when they are not aggressively feeding. Depending on the depth of the water, I will fish these flies using a regular bass leader or add a sink tip with a short leader to the end of my bass line. Fished with a slow retrieve, this fly is a good pattern for enticing a reluctant bass to eat. Note that all of the Woolly Buggers in this box are basically olive and black. It seems to be the only color combination needed.

Attractor baitfish patterns can also be tied with these materials. Chartreuse with a white belly and black with a purple belly are just a couple of the patterns that will attract bass as well as other warmwater fish.

Sinking tips that can be attached to a floating fly line are a versatile tackle item that can be used with swimming flies. With the different sink rates, flies can be fished at various depths up to about 12 feet. These tips are marketed for Spey fishing; however, they also work well with one-handed bass rods and lines. Normally the tips are 10 feet long. They can be used as a single tip, but if greater depth is needed, you can use two tips together with the tip that has the faster sink rate at the end. With the use of sink tips, the depth that a fly is fished can be controlled with a fair amount of precision.

JIGGING FLIES

Jigging flies are patterns that are bounced along the bottom of a lake. These patterns will have a tungsten cone (or some other metal cone), lead eyes tied at the eye of the hook, or some other type of weight added to the fly. One of the most common materials for tying these jig-type flies is a Zonker strip

Weighted jig flies are the best flies for searching out a big largemouth bass. A 9- or 10-weight rod is best suited to casting these heavy flies. Also, these patterns need to be fished using a floating line in order to be able to jig the fly vertically. I use a leader that has a tip that measures out to 16-pound test and then add 4 to 5 feet of 16-pound tippet. Do not attempt to make an overhead cast with this setup but instead use a sidearm cast. Additionally, keep the cast short and directed at a nearby target.

of rabbit (or pine squirrel if tying small jig flies). Some crayfish patterns also fall into this category depending on how they are tied.

Of all the types of bass flies considered here, this one is definitely the most difficult to cast. Use a sidearm cast for these flies, as they are heavy, especially when wet. Casting these flies is not like casting dry flies, and getting hit with one is painful. An overhead cast creates too much momentum, and the fly hits the water with too much force. Sidearm casts control the presentation and deliver the fly more softly. Casting sidearm and pitching a hare jig-style fly to the open pockets along the edges of lily pads or other cover can be very effective at attracting bass, especially if topwater flies are not working.

I use at least an 8-weight rod with a floating bass taper line for these flies. Nine-foot bass leaders that are 12- to 16-pound test and corresponding tippet of the same size are best. Tippet choice depends on the water being fished; heavier tippets are used in heavier cover. Adjust the length of the tippet to the depth of water that is being fished, with longer tippets for deeper water. These flies will work in water depths of up 10 feet; however, if trying to fish deeper than that, the length of leader needed makes casting really difficult. Unlike swimming flies, this style of fly does not work well when used in combination with sink tips. The floating fly line is what gives the fly the jigging action that makes these patterns so effective.

There are two basic retrieves when using this fly. The first is to hop the fly along the bottom, simulating a crayfish. This is a slow retrieve and is great for keeping the fly in front of the fish. Swimming one of these jig flies, simulating a baitfish, is also effective.

Choose colors according to the type of creature you are trying to imitate. Olives, browns, and oranges are best for crayfish, and olive/yellow and gray/white are best for baitfish. Chartreuse is always a good choice, as the fish can easily see it, especially if the water is turbid.

Boat control is important for all fishing but especially for fishing jig patterns. The most productive method is to cast to an identifiable piece of cover, allow the fly to sink, then begin a slow retrieve. This process takes time, so the boat must remain in a stable position to allow this to happen.

A STRATEGY FOR A DAY OF FISHING

I typically start my day by fishing the eastern side of the lake in the morning. If there are large trees lining the lake, I am fishing into the shade. I think

The olive coloration plus the black splotches along the lateral line are distinct characteristics of the largemouth bass.

Early morning mist indicates that the air temperature is cooler than the water temperature. This is a common phenomenon in the Great Lakes State.

the fish are more aggressive in the shade, plus I am less observable to the fish by being in the shade myself. If left undisturbed, bass, like trout, can be found in extremely shallow water. It is not unusual that some of the largest bass I catch in the summer come from water that is 3 feet deep or less.

As the day progresses, I continue to fish the shaded side of cover. Regardless of whether it's a dock, boat, lily pads, or weed line, the shaded side of the cover should always be the side you focus your efforts on. Brighter days will usually mean that the fish will hold tighter to the cover. This means you need to cast closer to the cover to entice the fish to eat your fly. On overcast days fish will normally cruise farther from the cover. Consequently, your casts can follow suit and be fished a little farther from the cover. When working shoreline cover, light and shadow are the indicators that dictate where you place your casts.

When the effectiveness of topwater flies begins to wane, I will then switch to fishing swimming flies in the deeper water just off the shoreline. If there are underwater weeds, I concentrate on the inside edge of the weed bank. I start by fishing the flies parallel to the shallow edge of the

weed line. If I am not catching fish, I will adjust the depth that I am fishing. I will continue to work deeper water until I find the bass again. Along with adjusting depth, I will also adjust how fast I retrieve the fly pattern.

If swimming flies are catching bass, I will continue to use them. However, if the bass are not in the chasing mood to hunt down a swimming fly, I will go to jigging fly patterns and fish them close to cover. If the day is bright, I'll fish close to the cover on the shaded side. Again, a caution: homeowners

Heading home after a day of fishing.

really do not like having their boats or docks hit. With fly fishing you can easily see the line rolling out, so errant casts can be corrected.

If you are fishing through a full day, as late afternoon rolls into evening, simply reversing this strategy is effective. I always enjoy casting topwater bass bugs into the summer twilight.

Find the shaded areas to fish in the morning.

This is a good example of natural cover and edges. Bass will hold in the shade under the lily pads.

While not as scenic as a natural shoreline, this shore provides man-made cover, structure, and edges. The pontoon boat and dock provide overhead cover and shade. The seawall has a hard edge that small panfish can get trapped up against, providing an easy meal for a foraging bass or pike.

Open-water cover like this mixture of reeds and lily pads provides a good target for jig-type flies. Fish them near the bottom along the edges of the emergent vegetation. The pockets in the weed beds are also good targets for jig-type flies; however, heavy tippets are needed to turn fish into the open water.

As the day progresses, search out overhead cover next to slightly deeper water. Given the time of day, a streamer pattern or Woolly Bugger fished parallel to the bed of lily pads is an effective presentation.

Nymph Fishing for Bass

Most anglers do not associate nymph fishing with largemouth bass. However, large nymphs can be a deadly way to pursue them, especially after a cold front has moved through during the summer and the fishing has slowed as a result. Dragonfly nymphs are commonly used for this purpose and are a good choice, although not the only possibility. Polly Rosborough's *Tying and Fishing the Fuzzy Nymphs* included patterns for several impressionistic nymphs. These flies are especially suited for fishing on large freestone rivers for trout. I have used them with good results on the Roaring Fork River, which has good populations of stoneflies, in Colorado.

While Rosborough did not intend his patterns to be used for bass and other warmwater fish, his impressionistic flies are well suited for that purpose. The two flies that work especially well are the Casual Dress Nymph and the Non-Descript Nymph. I have modified the tying of these two patterns to take advantage of modern tying materials. The Casual Dress Nymph can be tied with dubbing that includes some form of sparkle material. Sparkle material like Micro Krystal Flash can be worked into both the collar and tail. It can also be dressed with dyed materials to include colors other than the gray muskrat of the original pattern such as olive or black. The Non-Descript Nymph is also a good pattern to tie in various colors. One of the modifications I have made to this fly is to add a small Flymen Fish-Mask (size 3 or 4, depending on the size of the fly) with eyes. This makes a great small fingerling baitfish pattern.

When a cold front cools both the water and fish activity, bass can still be caught with a slowly fished nymph.

Largemouth bass rarely see large nymph patterns like these and will readily take them. Having a selection of large nymphs is essential for days when the bass are not in an aggressive mood. I normally add 4 to 5 feet of tippet to my regular bass leader when fishing nymphs. If fishing deeper, you can use a sinking VersiLeader of the appropriate sink rate.

Another great fly pattern for this purpose is Dave Whitlock's Red Fox Squirrel Hair Nymph. I tie the soft-hackle version of the fly and also will add some sparkle to the dubbing.

All of these flies are tied on a 3X or 4X long nymph hook such as the Mustad 9671 or Mustad 9672. Other hooks suitable for these patterns are the TMC 5262 and TMC 5263. I will weight the hook with a lead-substitute wire before tying these flies. These three flies will probably be all the large nymph patterns that need to be carried; however, any large soft-hackle fly patterns will also work.

To fish these flies in shallow water that is less than 6 feet deep, just use a floating line along with a leader that is from 9 to 12 feet long. For this, fishing a standard trout leader with a 3X tip works well. To the leader tip, add 20 to 30 inches of 3X or 4X tippet. For fishing deeper, add a sink tip to the end of the floating fly and then a shorter leader. Pick a sink tip that will be most efficient at getting the fly to the proper depth.

Finally, fish these flies slow. Retrieve the fly just fast enough to keep it from hitting and catching on the bottom of the lake. Remember, as with all fishing, keep the rod tip low to the water to effect a straight-line connection to the fly. This will facilitate detecting even the slightest hits to the fly. This is the last tactic I will use on slow fishing days. It is a rare day that a bass or other panfish can resist a slowly retrieved nymph pattern. ∎

6

Smallmouth Bass

Among anglers it is simply known as the incomparable smallmouth bass, the glorious bronzeback . . . a black bass with the grace, fight, cunning and intelligence of a trout.

—Lefty Kreh, *Lefty's Little Library of Fly Fishing: Fly Fishing for Bass*

It was a precarious walk down the bank to get to the river. The climb down, though not long, was steep. The river here was probably about 30 vertical feet from the top of the bank. In many places it was a truly vertical bank; however, this path, though steep, had a switchback halfway down the bank as it descended to the river. This river does not appear on any list of "Must Fish Places" and most fly fishers would simply drive past it without a second glance. I did for many years.

There are a couple of elements in the Great Lakes region that divide rivers into those that will sustain a trout population and those that will not. I grew up in Midland County in Michigan, and always found it curious that there were no natural lakes in the county. The two lakes that were in Midland County were impoundments of the Tittabawassa River, and they disappeared in May 2020. With already high lake levels from an unusually wet spring, the dam failed on the upper lake during a rainstorm that added 7 inches

A smallmouth bass from a western impoundment. PHOTO BY FRANK WHISPELL AT WWW.FFFADVENTURES.COM

of water to the area. Like dominoes, this caused the lower lake dam to be breached as well. Now there are no lakes in Midland County.

However, on the northwest corner of Midland County is Clare County. Clare County is home to probably over 100 lakes, with at least 18 of them having public access. For the most part these lakes are all natural. Lake George, my home lake, is spring-fed and drains into the headwaters of the Muskegon River. Also, there are three river systems that cut through the county: the Tobacco River, the Cedar River, and the Muskegon River. The Tobacco and Cedar have populations of trout.

Midland County is largely flat with mostly clay soils. This is why my river that I was about to fish had such a precarious entry down the bank. Over time the river cut through the clay soils until it hit a harder substrate. As you travel northwest out of Midland County into Clare County, the soils change from clay to sand and gravel. Additionally, you leave the flat country and head into the beginning of the hilly country of the northern half of the Lower Peninsula of Michigan.

The difference between these two areas is that Clare County and the counties to the north have the two elements that are needed both for the existence of rivers that sustain trout and for natural spring-fed lakes to be present. These elements are hills, plus sand and gravel soils. With these two elements, the winter snows as well as rain the rest of the year will percolate down into the soils and then come out at the lower locations, the rivers and lakes, as springs.

From Clare County north to the Straits of Mackinaw the soils are largely composed of sand and gravel, which were deposited by the period of glaciation over 10,000 years ago. Michigan was the only state that was completely covered with ice, and the many lakes, beaches, and sand and gravel soils are evidence of its presence. In the northern part of the state, many of the lakes have a northwest to southeast orientation, including Lake George, which corresponds to the predominant direction of the ice flows, further evidence of the effect of glaciation here.

Still, even though Midland County and other parts of the state with similar clays soils and flat terrain do not have the elements needed to sustain trout, there are still angling opportunities here. The river I was walking into provided one of those opportunities. This portion of the river was populated primarily by smallmouth bass, carp, and assorted panfish. I was here for the smallmouth bass.

ABOUT SMALLMOUTH BASS

Both largemouth bass and smallmouth bass belong to the sunfish family. Though similar in many ways, they are different species and have slightly different traits and habits. Smallmouth bass prefer cooler waters compared to largemouth bass. An ideal temperature range for them is from 60 to 75 degrees.

In addition to the bronze coloring, smallmouth bass also have red eyes. PHOTO BY FRANK WHISPELL AT WWW.FFFADVENTURES.COM

This range has a large intersection of temperatures that trout prefer as well. Perhaps this is why so many river systems have trout in the cooler headwaters and smallmouth bass in the lower and somewhat warmer stretches of the river.

As the name implies, smallmouth bass have smaller mouths than largemouth bass. If the mouth is closed the back of the jaw will be directly under the eye, rather than behind it as with largemouth bass. They will normally have a bronze color, whereas largemouth will be more olive colored. Vertical bars of color will also be present, although as they get larger, the bars become less noticeable. Like rock bass, they also have red eyes.

Smallmouth bass go through the same seasonal cycle as largemouth bass—pre-spawn, spawn, post-spawn, summer peak, and fall—it just occurs at a slightly lower water temperature and a little earlier in the year. They will begin spawning in the spring (or early summer) when water temperatures hit the 60- to 65-degree mark. Spawning in water that is up to about 12 feet deep, the ideal bottom is composed primarily of rock and gravel. The male bass will sweep a circle nest where sunlight is present but sometimes also next to a boulder or log. After the nest is ready, sometimes several females will lay eggs in the same nest before the male fertilizes them. Just as with largemouth bass, the male smallmouth will guard the nest, although the fry will leave the nest quickly after hatching.

Smallmouth bass will orient to different cover than largemouth bass. Smallmouth prefer rock and gravel areas, while largemouth tend to orient around softer sand and gravel bottoms and vegetation. If fishing for smallmouth bass in lakes, many of the weather conditions that affect largemouth behavior will also affect smallmouth behavior. However, in a river environment, cold fronts do not have the same impact as they do on lakes. The main environmental impact that affects river fishing is changes to the flows. Rivers, like the one described above, that are in areas with primarily clay soil can change flow rates very quickly. In this type of river, if a thunderstorm comes through with enough rain, flows can change from as little as 125 cfs (cubic feet per second) to as much as 2,500 cfs in a matter of a few hours, then the river will return to 125 cfs a day or two later. All the rain literally runs off of the clay soils and into the river. By contrast, the Au Sable River, in the northern Lower Peninsula, which flows through sand and gravel soils, is one of the most stable river systems in the world, and the flows will hardly change with a similar storm.

A large, sudden change in flow may make a river unsafe to fish, but along with a change in water levels, a change in temperature will happen as well. All fish are cold-blooded, meaning their body temperature is the same as the water temperature. However, a lower body temperature also means their metabolism slows down. Normally, higher flows will lower the temperature of the river. Conversely, a drop in the river level will not usually have a negative effect on fishing, as the temperatures are not dropping. If the temperature of the water is rising, it can improve the fishing because the warmer water temperatures will mean the fish will have a higher metabolic rate. Stable river flows create the best fishing for smallmouth bass.

TACKLE FOR SMALLMOUTH BASS

A good rod for river fishing for smallmouth is a 7-weight rod that is 9 feet long. When paired with an appropriate line, this rod will not only handle the flies being fished but also be long enough to effectively fish some of the tactics that are unique to a river. A floating bass taper line will be needed to cast the flies. If wading a river, rarely will you need to add a sink tip to the floating line. Only for larger rivers that perhaps are being fished from a boat will a sink tip be necessary. Using bass leaders and tippets will be the norm. Rarely in rivers is there a need to fish a tippet heavier than 12-pound test. For more information on tackle, see chapter 3, "Modern Tackle for Warmwater Fly Fishing."

FLIES AND TACTICS FOR SMALLMOUTH

Most of the flies that are fished for largemouth bass will work for smallmouth bass as well. If fishing for smallmouth in lakes, many of the tactics for large-mouth will apply. If river fishing, many of the same methods used for trout will work for smallmouth bass.

When fishing topwater poppers or bass bugs in a river, it is best to work the fly across the currents. Working a popper out from a bank that holds fish is

Many rivers provide angling opportunities for smallmouth bass.

This box contains a variety of fly patterns for river fishing. I carry it with a small chest pack and a few tackle items. The crayfish imitations in the upper right side are the flies I use most often. When crayfish are present in a river system, they will normally be the main food source for the smallmouth bass. I use these same patterns for carp, which commonly share the river system with the smallmouth. Carp will also readily eat crayfish when available.

effective. Cast and mend so as to have the popper move in a direction that is across the currents rather than upstream or downstream. Topwater fishing is best in the early morning. Just as in bass lakes, try to work the areas that are shaded.

Dry flies can also be fished. Terrestrials like hopper patterns or large dry flies like an Irresistible work well. Fish dry flies the same way you would fish them for trout. Cast and mend the line in order to get a drag-free float. Unlike poppers, the fly should not be skated across the currents. Dry flies work well later in the day as insects like grasshoppers, crickets, and bees become more active.

Weighted Woolly Buggers and streamers also work. If the water is deep enough, this is where adding a sink tip could mean the difference between catching or not catching fish. Standard streamer fishing tactics apply, where the fly is cast across or across and slightly downstream. Mend the line so the fly will swim across the currents. Allow the fly to completely sweep across

Connecting a crayfish fly with a twist clip allows it to drop tail (at the eye of the hook) first. The claws extend upward, and when jigged across a rocky bank of a lake or through a run in a river, it is a presentation that smallmouth bass find hard to resist.

Crayfish, when present, will be a prime forage item for smallmouth bass.

Rivers are prime locations to find smallmouth bass. PHOTO BY FRANK WHISPELL AT WWW.FFFADVENTURES.COM

the currents until the line fully straightens out downstream. Keep the fly in the water for a moment, because if a bass has followed the fly across the currents, this may be the point where it will take it. This tactic works well by starting at the head of a deeper run and taking a step downstream after each cast. If the fly does not hit the bottom and there are no strikes, consider changing to a heavier sinking tip or a heavier fly in order to fish the fly deeper through the run.

There is a fly of note that needs to be mentioned here. If crayfish are present in the river system that is being fished, a good crayfish pattern is sometimes all that is needed. In 2016, western Colorado artist and fly tier Bill Fenstermaker first showed me his crayfish pattern at the Eastern Rocky Mountain Council Regional Expo in Glenwood Springs, Colorado. It is tied on a Gamakatsu 60409 jig hook. A tungsten cone is threaded onto the hook, so it butts up against the 90-degree bend before the eye of the hook. Lead or substitute lead wire is then wrapped around the hook shank just behind the cone. This gives the fly the weight that is needed to get to the bottom quickly. The body is then tied in reverse, so the head and claws are at the bend of the hook and the tail is at the eye of the hook. Other crawfish patterns are tied in this manner as well. When crawfish feel threatened, they will drop their tails to the bottom and extend their claws upward in a defensive posture. Fly patterns that mimic this behavior are very effective.

I normally use a standard bass leader with a 12-pound test tip. To the tip of the leader add a 2-foot section of 12-pound test Amnesia monofilament that is either chartreuse or fluorescent red. Finally add a section of 12-pound test tippet. The Amnesia monofilament acts as a sighter material. Fishing this set-up is like Euro nymphing in that it is fished on a tight line. Cast it up and across the current, then lift the rod tip to tighten the line and follow the fly with the rod tip as it drifts down through the run. Jig the pattern up and down. Keep a tight-enough line to feel the fly coming in contact with the bottom of the river. If done properly, it is easy to detect a take when a fish grabs the fly. The sighter material will also twitch or move when a fish takes the fly less aggressively. On the river that I described above, this is literally about the only tactic that I use, although I will occasionally try something different just to see if I can catch fish. There is the added advantage that if you wade up to a suspected carp hole, the same setup and tactics can be used there as well. Large nymph patterns can be fished with this same setup. The clip at the end of the leader makes it easy to switch flies and experiment with different patterns.

Locate a good smallmouth bass river and give this fishing a try. If you do, you will probably understand why so many anglers will fish for river smallmouth bass as a substitute for trout—or just to have a great fishing experience in and of itself.

7

Panfish

*It is impossible to grow weary of a sport that is never
the same on any two days of the year.*

—Theodore Gordon, 1914

As the morning progressed, the bass fishing slowed. It had been about 30 minutes since I had moved a fish with my topwater bass bug. Rather than fishing the deeper water for bass, I pulled out my 4-weight rod and began fishing a small popper for the many panfish that populated the shallow water. I began catching fish almost immediately, mainly a mix of bluegills and pumpkinseeds. I worked my way down the shoreline, steadily catching fish for the next hour. Most of them were smaller but mixed in were some of the larger ones plus an occasional rock bass. This is an enjoyable way to use a fly rod.

Panfish are often considered a great way to begin fly fishing. They are not as selective as trout; with trout you will normally need at least a rudimentary understanding of aquatic insects and the flies that imitate them. Additionally, if you are fishing a river, you will need to understand how to get drag-free floats with the fly as it is presented to the fish. With bass and other warmwater predator fish, being able to cast larger rods as well as larger wind-resistant flies is essential, and gaining casting expertise takes time and practice. With panfish the fly selection is much simpler and 50- to 60-foot casts are not necessary. As a beginning angler, I honed both my fishing and casting skills on panfish.

Ponds with abundant food will produce outsize bluegill.

However, panfish should not be thought of as only a species to pursue as a novice. Time spent pursuing panfish should be a part of every angler's season. They are a worthy game species and while catching the smaller ones can be quite easy at times, catching the larger panfish takes understanding and skill.

ABOUT BLUEGILL

Bluegill, like bass, are part of the sunfish family. The most common panfish encountered are bluegill, which are sometimes referred to as bream or brim. They share the same water and will follow a similar seasonal pattern as that of largemouth bass. Most panfish will become a little more active during the spring warm-up and pre-spawn period than bass will; however, they will normally spawn after bass have spawned. When water temperatures reach the high 60s to 70-degree mark most bluegill will begin spawning, sweeping out nests in hard sand and gravel bottoms. These nests are a familiar sight in many lakes. Like the largemouth bass, the males will guard the nest after the spawn until the eggs hatch and the fingerling bluegill leave the nest.

Small bluegill will school, and often the schools will contain other panfish like crappie, rock bass, and other sunfishes. These panfish schools are

An abundance of aquatic insect life creates a prime fishery for this Pelican Lake bluegill.

the main source of food for the predator fish that share the same water. Unlike largemouth bass, bluegill will continue spawning into August. It is not unusual to find active nests throughout most of the summer. As the summer progresses, when you find smaller bluegill in the shallow water, it is not unusual to find the larger, more mature bluegill in the deeper water just out from the smaller fish.

ABOUT PUMPKINSEED

The pumpkinseed is another very common panfish found in many lakes. Many of the characteristics of the bluegill are shared with the pumpkinseed. However, they will begin to spawn earlier that the bluegill, with spawning starting when the water gets into the mid-60s. Like bluegill, they will continue to spawn, and active nests can be found throughout the summer.

Although bluegill can grow larger, pumpkinseeds will still get large enough in most lakes to be of interest to a fly angler. Often pumpkinseed, bluegill, and other species of fish will be intermixed along a shoreline, so targeting a specific species is difficult. The pumpkinseed is one of the most colorful of the panfish, with turquoise, orange, and gold sides.

All fish feed more aggressively in shaded water, which is where this pumpkinseed was caught.

Crappie will sometimes come up to the surface to eat a topwater bass bug; however, they are normally caught when fishing small streamers in deeper water.

ABOUT CRAPPIE

The black crappie and white crappie are both part of the sunfish family. Crappie (pronounced "crop-ee") will school as adult fish and tend to be more predatory, preferring small baitfish to other food sources. They will spawn in the spring when the water temperatures are in the mid-60s. Fishing for them pre-spawn can be fruitful, as crappie will remain active all winter so are more active in spring than most other sunfishes. Spring is considered the prime time to fish for them, but I have caught them all summer long, although in fewer numbers than most of the other sunfishes included here.

While small streamer patterns are good flies to target crappie, they have a bigger mouth than bluegill or pumpkinseed. They will take larger flies because of that. I have caught them with topwater bass bugs while fishing for bass.

ABOUT ROCK BASS

The rock bass, another member of the sunfish family, is perhaps the least appreciated of this group. They tend to be more of a monochrome bronze or olive color, though they do have a more startling red eye. Like crappie, they have bigger mouths than other sunfish and are readily caught with larger flies. Being part of the sunfish family, they too will spawn in the spring when water temperatures get into the high 60s and 70s.

Rock bass can be identified by their olive-bronze color and red eyes like those of a smallmouth bass.

I will often catch rock bass while fishing topwater bass bugs for large-mouth bass. The fact that their mouths are larger than other sunfishes allows them to eat larger flies. However, I usually know when I have hooked a rock bass as opposed to a largemouth bass: like all of the other panfish they rarely jump, whereas one of the first things that a largemouth will do is jump.

TACKLE FOR PANFISH

Fly rods can range in size from a 2-weight up to a 6-weight for panfish. If conditions are right, I have a 2-weight rod that is 6 feet long that I will use for panfish. This is an opportunity to really have some fun with very light-weight fly tackle. Since the flies will also be about the same size as trout flies, the same lines, leaders, and tippets that are used for trout fishing can be used here.

Using these lightweight rods to sight-fish to large bluegill or pumpkin-seed never gets old. Plus, the occasional largemouth bass may swim by, and enticing a large fish like that on a small rod and fly is an extra-special experience. For more information on tackle, see chapter 3, "Modern Tackle for Warmwater Fly Fishing."

With a gust of wind, this wasp would become easy pickings for a waiting panfish.

This tri-colored bumblebee is pollinating a pickerel weed. Pickerel weed is a common flowering water plant around many lakes.

FLIES FOR PANFISH

Bluegill and other sunfishes are not necessarily selective feeders but rather more opportunistic feeders. During a large aquatic hatch the bigger fish can become a little more selective, but not to the extent that a trout will selectively feed on a specific insect or at times a specific stage of an aquatic insect's emergence. So, a fly selection for these fish does not need look like one for trout but rather can be more suggestive.

One of the easiest places to start in your panfish fly selection is with simple soft-hackle patterns. Two productive ones are a muskrat gray dubbed body with a grizzly hackle and a hare's ear dubbed body with a partridge hackle. I will normally weight these patterns with a few turns of a lead-substitute wire on the hook shank and tie the fly over the weight. You could also begin the fly with a bead at the front to give it some weight. Soft hackles are where you can get creative and use different colors of dubbing, especially if the dubbing contains some sparkle material. Any traditional soft hackle for trout will work for panfish as well. I carry these flies from a size 10 to a size 14. Bluegill and pumpkinseed will take smaller flies, but the small size of their mouth along with the fact that, like largemouth bass, they suck in the fly pattern, they then become difficult to unhook. Have a pair of hemostats on hand to unhook the fly and release the fish.

Along with soft-hackle flies, any nymphs will work as well. One of the patterns I carry is the A.P. Nymph. This pattern was designed by Andre Puyans in the 1980s and has been a steadfast fly in my boxes for both trout and panfish. A modern version of this fly contains a bead at the head of the pattern. For panfish the standard pattern is fine, as the bead is more applicable to river fishing for trout. I tie this fly in three different versions: muskrat gray, olive-brown, and pheasant tail as well as black. It is normally tied on a 1X long nymph hook; however, it can be tied on a 2X or 3X long nymph hook as a good imitation for a damselfly nymph in olive-brown.

This is one of my go-to boxes for panfish. It includes soft hackles, which are among my most effective flies. Also note the bead-head flies, which often function as the dropper in the popper-dropper combination (see the "The Popper-Dropper Combination" sidebar). This box also includes a few bucktail streamers and other assorted nymphs.

This box holds assorted small poppers for panfish along with some rubber spiders that are mandatory patterns for these fish. Using these patterns with a small light rod is a great way to go. Also note the bee and wasp patterns in this box. Flies like these along with other terrestrial patterns are also effective for panfish.

No panfish fly collection would be complete without some foam or rubber spiders. These simple flies float well and with the rubber legs moving about are very enticing for bluegill and other panfish. Along with rubber spiders, I also have a collection of small poppers. This is a collection where rather than tying them, I have purchased them. Although I have made and tied balsa wood poppers, gluing the popper body to the hook shank and then painting it is an involved process. While I enjoy doing this sometimes, it is easier to just purchase a few. If creating your own hard-body popper-style flies interests you, Steve Schweitzer's book *Designing Poppers, Sliders & Divers* is the absolute go-to reference for these flies.

Small streamers and bucktails are also useful, especially when fishing for crappie. Here too you can use your imagination in creating them. Traditional patterns work, but by creating a body with some flash and then adding either a bucktail or feather wing, simple patterns can be tied that will cover most situations. Simple bucktails with wings that are either white, yellow, chartreuse, and olive are all effective.

Dry flies are useful as well. I have some standard fly patterns such as the Adams as well as some high-floating western dry flies like the Irresistible. With all these flies you can be creative with the colors you choose. Included in this list would be terrestrial flies like hoppers, crickets, ants, and other insects that are present along the shoreline of where you fish.

Damselflies and dragonflies are present in every lake I have ever fished and are a familiar food source for panfish. Adult damselfly and dragonfly patterns as well as a nymph for both are reliable patterns to include in a panfish fly box.

This is, at best, a beginning list. Many other fly patterns will work for panfish; however, don't overcomplicate your selection of fly patterns here. Panfish are not trout—they do not need as wide a range of fly patterns.

When grasshoppers make an appearance, all fish find them hard to resist. Hopper patterns should be included in your warmwater boxes as well as your trout boxes.

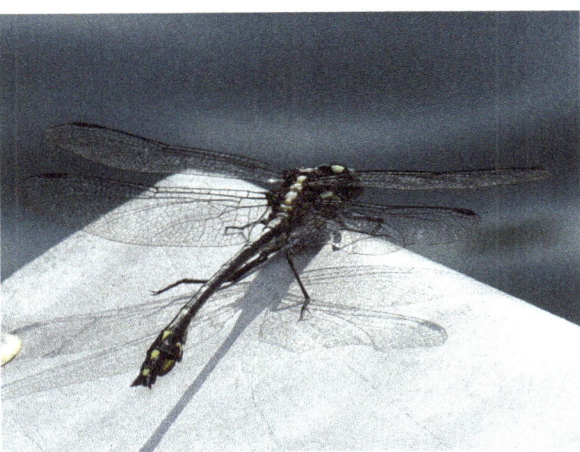

Dragonflies are common around most lakes, so while they are strong fliers, both panfish and bass are on the alert for when they end up in the water.

TACTICS FOR PANFISH

Perhaps the most enjoyable way to fish for bluegill and other panfish is with topwater poppers. All of these fish to greater or lesser degrees are very topwater conscious, and fishing poppers is an enjoyable way to catch them. When I am fishing for trout, if a trout rises to a fly but then does not eat it, one of my first adjustments is to pick the same fly in the next smaller size; this is often the only adjustment needed to catch trout. The same holds true for panfish.

I tie deer hair bugs on size 10 and 12 TMC 8089 hooks. These are the two smallest sizes for this model of stinger hook and are a good size for deer hair bugs for panfish. Smaller panfish can strike at these flies but they are too big for them, so only the larger panfish can eat them. If I keep getting hits to the fly but not hookups, then just as I do with trout fishing, I will downsize to a smaller popper. Normally that is all it takes to start getting hookups, but if all of the fish are small, I might try a popper-dropper combination (see the chapter 7 sidebar, "The Popper-Dropper Combination").

Another way to search for bigger panfish is to fish the deeper water just out from where the smaller panfish are. If the water is not more than 6 feet deep, tie a longer piece of tippet to your leader and use a lightly weighted nymph or soft hackle. When you cast the fly, drop the rod tip to get a straight-line connection to the fly and then tighten up slightly. Use the countdown method

For many kids, panfish are the first fish caught on a fly rod.

While it could just as easily have been a bass eating this small deer hair popper, this large bluegill decided it looked like a good meal also.

to let the fly sink. Be patient and let the fly get near the bottom, then begin a slow retrieve. You could also fish larger nymphs, as they will attract the larger panfish with the added bonus of maybe attracting a largemouth bass (see the chapter 5 sidebar "Nymph Fishing for Bass").

An effective way to get deeper or have the fly drop down faster is to add a sink tip to the end of the fly line, still using the countdown method to reach the desired depth. With the use of sink tips, 10- to 12-foot depths can be effectively fished. However, fly rods that are lighter than a 5-weight will have difficulty casting an added sink tip.

Angling for panfish, especially the larger ones, is a sport worthy of pursuit for any fly fisher.

The Popper-Dropper Combination

The hopper-copper-dropper combination is a well-known and popular method for fishing for trout. It involves a high-floating dry fly (in the West this is usually a foam hopper pattern of some sort), with a bead-head nymph tied to the back of the fly with 18 to 24 inches of tippet. The combination got its name from using a hopper fly pattern along with a Copper John nymph as the dropper, although any bead-head nymph can be used.

In his book *Rivers of Sand*, Josh Greenberg writes of fishing droppers, using high-floating dry flies with a small glittery nymph as the dropper. The dropper nymph is always attached with fluorocarbon leader material. I have put this tactic to good use in Colorado, especially during the Green Drake hatch. Using a large Green Drake dry fly with a bead-head Brown Hackle Peacock nymph as a dropper has always been a winning combination on the Frying Pan River. One of the more subtle tricks with this method is to attach an additional gold tungsten bead to the fluorocarbon tippet material between the dry fly and the nymph. Let the bead freely slide along the leader material between the two flies. When you cast, the bead will naturally slide down to

Hand-size panfish like this sunfish will be attracted to a popping bass bug but will instead opt for a small nymph dropper fly.

the nymph. I have used this dry-dropper combination both on western rivers and on rivers like the Au Sable in Michigan with good success.

The popper-dropper combination is as effective for panfish as the hopper-copper-dropper combination is for trout. With this method I can use a deer hair bass bug, although foam or balsa poppers also work well, as they create a distinctive pop in the water as they are retrieved. While fluorocarbon leader material can be used to attach the dropper, with panfish it is not necessary. Panfish are not as leader-shy as trout. Just as with trout, nymphs with a fair amount of glitter work well to attract the attention of the fish. Attach the dropper at least 24 inches behind the popper with 3X or 4X tippet material. Also, just as with a trout setup, an additional bead can be added between the popper and the nymph. However, since you will probably be fishing this method in a lake, the additional bead is not as necessary as it is when fishing a river.

The popping sound the topwater fly makes as it is retrieved will attract the panfish, and when they investigate, they will see the nymph and take it instead. The floating fly acts as a strike indicator. Set the hook when you detect any unnatural movement with the topwater popper. Sometimes the fish will grab the nymph while the fly is being retrieved. When this happens, the take can be felt, although it is more subdued than when the topwater fly is grabbed. This occurs because line, leader, popper, and nymph are not in a straight line; there is normally an angle of line between the topwater fly and the nymph. Panfish will sometimes grab the nymph when the retrieve has stopped and it comes to rest in the water. The nymph will still have some movement that will attract the fish. When this happens, it is not unusual to see the topwater popper disappear from the surface of the water.

Just as with trout fishing, it is important to remember that you are fishing a two-fly rig. I have had the second hook impale my finger or hand as the fish was struggling to be released. It is really an unpleasant experience! ∎

Tenkara and Panfish

My grandfather led the way, as we stepped into the tackle and bait shop in the village of Lake George. This was always one of my favorite places to go. As I stepped in, the cool air would always greet me. The building was stone and concrete, so it was always cooler than the outside summer air. Upon entering, I was greeted with the sound of running water of the baitfish and crayfish tanks. The crayfish tanks were always the ones that fascinated me. I was drawn to them but also a bit terrified. I was probably only eight or nine years old at the time. There was also, of course, the refrigerator with the night crawlers in it. This was often our destination, as getting worms for bluegill fishing was part of our fishing plan.

Hanging on the walls of the shop were assorted lures, lines, and floats as well as various other tackle items, including rods and reels. There was always some time given to window-shopping, but we already had our rods and reels and unlike today it was a more frugal time. Also present were the bamboo poles. We already owned several, and bamboo poles were a staple of summer fishing for me as a boy. A line tied to the end with a bobber or float along with a hook and the purchased box of worms was all that was needed. Bluegill dinners were not uncommon as I was growing up.

Cane poles of old were not the precursor to tenkara rods, since tenkara rods were developed in Japan. However, both systems have some commonalities.

Tenkara rods are commonly used for trout; however, they also are an enjoyable way to fish for panfish.

Aside from this being an iconic image of boyhood fishing, there was elegance to its simplicity—an elegance that is captured in today's tenkara fishing. Tenkara originated in Japan as a way of fishing for the mountain stream trout and was originally used by commercial fishermen as a simple way to catch those trout.

It is natural to associate the bamboo pole of my youth with the tenkara rod of today, but the tenkara rods of today are extremely light and efficient fishing tools. The bobbers and worms are replaced by leaders, tippet, and flies. I do fish for trout with my tenkara rod, but I never fail to pull out the rod for some summer panfish angling. I use the same flies that I use with my fly rod, but the close and visual aspect of using the tenkara rod is enjoyable and a nice nod to my boyhood fishing.

There are lines that are designed for tenkara rods, but often I will simply use a 9-foot monofilament or fluorocarbon leader with up to 3 or 4 feet of matching tippet. I will match the tippet material to the leader material. Normally, I will use fluorocarbon if I am using sinking flies like soft hackles or nymphs. The action is imparted to the fly by twitching the tip of the rod. Using the length and flexibility of the rod to swim the fly along, I will watch the fly and then set the hook when a fish grabs it. This is very visual fishing.

I mostly do this while fishing off a dock, but for a change of pace, tenkara rods can be used out of a boat or when wading. Owning a tenkara rod is economical since, like the cane pole, the rod does not use a reel. Add one to your tackle bag and perhaps, like me, you will come to enjoy this simple and elegant method of catching fish. ■

8

Carp

You see, the thing is, carp are not suckers—not literally nor figuratively. They're wary. They're cunning. They're the ultimate piscine "survivalists."
—Kirk Deeter, *The Orvis Guide to Fly Fishing for Carp*

When my longtime fishing friend Mike called me in late June with an invitation to come to his home in southern Michigan, I knew I was in for a unique fishing experience. The next day I pulled into the parking lot for the boat ramp where we had agreed to meet. We were fishing the backwaters of the St. Joseph River near his home in Three Rivers. The mulberries along the banks were ripe and dropping into the river. This part of the river was quite wide and had a barely perceptible current. There were several other boats out in the river, including a few pontoon boats. Some were people just cruising around, but there were other fishermen as well. Most of them were anchored out in the middle of the river and fishing from there.

After launching Mike's canoe, we paddled along the bank looking for mulberry bushes with ripe fruit. Although there were other fishermen on this stretch of the river, we were the only ones who were fishing the bank. We paddled only a short way before we came to the first mulberry bush that Mike had identified earlier. I was fishing a Mulberry fly that was tied on a size 10 TMC 8089 hook and consisted of wrapped

Western reservoirs are good locations to find carp. PHOTO BY FRANK WHISPELL AT WWW.FFFADVENTURES.COM

Frank Whispell with a large western common carp. PHOTO BY FRANK WHISPELL AT WWW.FFFADVENTURES.COM

chenille that was a dark purple and treated with a coating of thinned goop. The fly was tied to a 20-pound fluorocarbon tippet.

When I made my first cast along the bank, the fly landed with a satisfying plop, much like the actual berries, and slowly sank. Almost immediately the tip of my fly line began to bend and I felt a pull heading for deeper water. I set the hook with some authority and a violent thrashing immediately ensued. The fish turned back to the bank and the cover, and the fight was over almost before it began. I stripped in my fly line to discover that the hook had straightened out and that was how I had lost him. This was my introduction to carp fishing.

As we continued along the bank, each mulberry bush would have several carp under them but at most we would hook perhaps two. However, normally each bush was only good for the first fish, then we would have to move on to the next bush. These were the strongest freshwater fish that I had ever caught, and many of them would run almost to the backing. Even with the drag set at the higher levels, the carp were still able to run out a fair amount of line.

After having hooked numerous fish, we worked our way a few hundred yard along the bank. A couple of fishermen, who were out in the middle of the backwater fishing from a pontoon boat, shouted across the water to see if we were catching smallmouth bass. When we replied that we were catching carp, they immediately dismissed us as pursuing something that was not worth their time. I was slightly incredulous that they had witnessed us hooking and fighting these powerful fish only to quickly shrug off the experience as something beneath them, all because of the species of fish involved.

Kirk Deeter, in his book *The Orvis Guide to Fly Fishing for Carp*, compared carp fishing to soccer. Much like the game of soccer, carp are a prized gamefish in much of the rest of the world; however, here in the United States they are rather easily dismissed as being beneath most anglers. While that analogy may not hold up today as well as it did almost a decade ago, there is still some truth in it. However, I know this to be true: many well-known fly anglers noted for other more "acceptable" fish species will pursue carp when an opportunity is presented. For myself, I approach an opportunity to fish for carp with a great deal of enthusiasm. The mulberry story above is very much the exception when it comes to carp fishing. I have discovered what so many other anglers have, that these are perhaps the most challenging freshwater fish to catch.

ABOUT CARP

This chapter focuses mainly on the common carp. As its name implies, it is the most widely distributed species of carp in North America. Unlike all the rest of the species described in this book, however, the carp is not considered native to North America. Carp were originally native to Asia; however, they were introduced into Europe centuries ago. So, for most purposes they are considered almost native to most of Europe. Izaak Walton includes a chapter on carp fishing in *The Compleat Angler*, describing the fish as "the Queen of Rivers." Walton recorded that carp had arrived as least 100 years before he was writing *The Compleat Angler*, which would place its introduction into England during the mid-1400s.

Along with brown trout, carp were introduced to the United States during the 1870s. Brown trout were introduced to the Pere Marquette River near Baldwin, Michigan, while carp were introduced in numerous places, both on the East Coast and the West Coast. Carp are very adaptable to a variety of water types and conditions, and it did not take long before they were present in numerous rivers, lakes, ponds, and reservoirs. Their adaptability may be one of the reasons they are held in such low regard.

Carp prefer temperate water conditions, with temperatures in the 60- to 70-degree range. However, unlike brown trout, they can survive in a variety of water types as well as a wide range of water temperatures. One of the

Beauty is in the eye of the beholder. The challenge of hooking up and their fighting qualities make carp worth pursuing. PHOTO BY FRANK WHISPELL AT WWW.FFFADVENTURES.COM

unique abilities of carp is that they can survive in warm stagnant water that has very low oxygen content. They have the ability to rise to the surface and gulp air in order to survive these conditions. Perhaps it is this quality that places this fish in such low esteem in the eyes of so many anglers. However, on the other side of the coin, carp thrive in very pristine conditions. The Great Lakes have an established carp population and, at the right times of year, provides a flats fishing experience very similar to fishing the saltwater flats for bonefish or redfish. Also, carp will spawn several times a year, with the female depositing up to 300,000 eggs each time. Carp are relatively longed lived and can survive under the right conditions for up to 50 years or more. Understanding all the advantages this fish has helps to understand why it is prevalent in so many waters.

Along with their unique abilities that make them so adaptable to so many conditions, carp also possess senses that make them particularly challenging to catch with a fly rod (or any type of fishing tackle for that matter). Carp have a heightened sense of smell and often feed by smell. While this may be an advantage if using fresh bait, it is a disadvantage for fly fishing, where our flies do not smell like anything edible. A simple way to help our flies smell more natural is to rub them around on the bottom mud, sand, or gravel of the water we are fishing. It's a small thing but does help eliminate some of the offensive odor of the fly. There are scents available that can be added to flies; however, this is perhaps a bridge too far for most fly fishers. I have experimented with scents on a few occasions, and you definitely do not want to return a fly to the box after doing so. Personally, I have decided that scents are not for me; however, each angler can determine their own course of action in that regard.

Carp also have acute hearing, with a lateral line as well as ears. A cautious approach is needed (this is true for all fish as well) to keep from alerting them to your presence. I have walked the banks of rivers and had carp take off while I was still some distance from them. Oftentimes once in position to fish for carp, it is wise to simply stand still and not fish for a while and just observe what the fish are doing (this is good advice for trout as well).

With their wide adaptability to various water types as well as their acute senses, this fish is probably one of the most challenging freshwater fish an angler can pursue. If you are able to catch carp on a regular basis, you can most certainly catch all of the other fish described in this book.

TACKLE FOR CARP

Selecting the right gear for carp fishing is a matter of personal choice; however, a few guidelines will help in getting started. When considering a rod to use, remember that longer rods give you greater leverage in casting a line than shorter rods but give up that leverage when fighting fish.

Carp are common in many warmwater rivers. PHOTO BY GARY ENGLISH.

One of my favorite rods for carp fishing in rivers is a 7-foot, 11-inch Sage Bluegill. This rod is considered a 6-weight. I like the shorter rod when it comes to setting the hook and fighting the fish. Carp will make long runs but more often will stop short of 100-yard-plus runs, especially in rivers. If I am using a sufficiently strong leader and tippet, I can exert considerable pressure on the fish with the shorter rod. I also use a couple of 9-foot, 6-weight rods. One of them is considered a saltwater rod and has a robust butt section, so it does a good job of fighting the fish. The other is a freshwater rod, but it is a very fast rod, so it has a pretty stout butt section as well. Besides casting, one of the advantages of the longer rod is that the extra length allows you to better manipulate the line on the water when fishing. If you are fishing larger bodies of water such as a flat in the Great Lakes or a large western reservoir, then probably a 9-foot, 8-weight rod is better suited to those conditions.

When it comes to selecting a reel, all of the reasons given in chapter 3 for large-arbor reels apply here. Having a good drag system on the reel is essential. Once on the Pine River in Michigan, I caught a carp that was in the 20-pound range. I was fishing a deep hole along the bank but standing on a wide, shallow sand flat that stretched behind me for about 50 yards or more. Once I hooked the fish and moved it up into the shallow water, it took off to the other side of the river. I started with the drag at about half of the pressure it would put on the line. Whenever the fish stopped, I would walk toward it, reeling in line as I went.

As this process of the fish running and my chasing it continued, I continued to advance the drag. In the final stages of the fight, I had the drag on the reel at the maximum setting. Still the carp would pull off about 10 to 12 feet of line in these violent head-shaking runs that felt like the rod was being jerked out of my hands. After about seven or eight of these runs, I was finally able to grab the fish by the tail. Without a good drag system on the reel, I am not sure I would have landed that fish.

I only use floating lines for carp, since this is primarily a sight-fishing game. Specialized weight-forward lines are best suited to this fishing. Lines like Rio's OutBound Short or Scientific Anglers' Amplitude Bass Bug are two lines that work well. Like fly rods, though, there are many other lines that will do the job, and finding one that works well for your fishing is largely a matter of personal choice.

When setting up a leader and tippet for carp, I will often tie my own starting with 30-pound Maxima Chameleon monofilament, then stepping down to 25-pound test, and to that I will add some 20-pound-test fluorocarbon for a total length of about 9 to 10 feet. I then add a twist clip to attach the fly. This leader works well for weighted flies. Bass leaders can also be used, matching the pound-test tip and tippet to the fishing conditions. For more information on tackle, see chapter 3, "Modern Tackle for Warmwater Fly Fishing."

Carp are some of the most powerful fish that can be caught in freshwater.

FLIES AND TACTICS FOR CARP

Carp are omnivores, eating both plants and animals. The biggest factor in choosing a fly is identifying what they are eating at the time. When the mulberries were ripe along the St. Joe River, that determined the fly of choice. The Pine River in Michigan has an abundant supply of crayfish, so using a crayfish pattern tends to be a starting point with those carp. I have rarely had to resort to any other fly there. With large western reservoirs, perhaps a dragonfly nymph might be called for. Since carp eat a wide range of food items, it really is a matter of observation to determine fly choice. It's always a good idea to sit and observe for a while before fishing, and with carp this is especially true.

As far as tactics for fishing to carp, the best advice is to first and foremost always work on a cautious approach to avoid alerting the fish to your presence. If the carp are feeding on some form of plant, like mulberries, matching the fly to that food item is important but so is presenting it to the fish in a manner consistent with how the natural food item is becoming available. This is really just good fishing advice for pursuing any species of fish.

If crayfish are present in a river system that has carp, that is a go-to food item that, like smallmouth bass, carp find hard to resist. In river systems that have both carp and smallmouth bass, the bass will occupy the places in the river with more current and perhaps a rocky or gravel bottom. The carp will be found in the slower water, especially if there are depressions in the river bottom. As described in chapter 6 on fishing for smallmouth bass, I will attach a crayfish imitation to my tippet with a twist clip. This allows the fly to freely hinge, and if the hook eye (tail of the crayfish pattern) is weighted, a very convincing presentation is possible. Crayfish, when feeling threatened, will drop tail down to the bottom with their claws extended upward. With a twist clip and properly tied crayfish pattern, this behavior can be accurately imitated. I always start at the head of the depression and, much like Euro nymphing, gently bounce the crayfish pattern on a tight line down into the deeper water. Along with crayfish imitations, other flies with sufficient weight can be fished on a tight line, including large bead-head nymphs.

If you can see the bottom of the hole and observe fish activity, often a strike can be detected visually. However, when the water is deep enough that the bottom can't be seen, the fly is fished by feeling the strike. Often, especially with big carp, the fly will simply come to a stop. If this happens, maintain a tight line and simply wait to see if you detect a swimming action. It's not unusual for it to take a few seconds for this to happen. If there is no swimming action, the fly has obviously come into contact with some wood or weeds (if present) or perhaps has become caught on rock.

However, if suddenly the line and fly begin to swim, get the line on the reel and hold on! In a river, if the fish runs out a lot of line, wade after the fish to keep it on a shorter line. (This is also true when trout fishing and using light tippets.) The longer the line being used to a fish, the more the current in the river will put pressure on the leader and tippet. Where it is safe to do so, wade after the fish to keep it on a short line. Of course, only wade in places where you are sure of a safe water depth. I have lost some very big fish because it became unsafe to wade after them. While losing the fish is frustrating, in the end I realize that no fish is worth getting into a dangerous situation in order to land it.

Probably the most challenging way to catch carp is out in a large flat, whether in a big western reservoir or in one of the Great Lakes. I will leave it to others to write about carp fishing in the Great Lakes, but as far as reservoirs are concerned, my best advice is to be patient and observe the carp's behavior. If the fish are just cruising around, it's probably best to just let them go and wait until they are passing by slowly or even better, tipping down and eating something off the bottom. If the bottom is somewhat soft, you will see mud being disturbed. After determining the direction the fish are moving, cast the fly a couple of feet ahead and let it settle to the bottom. As the fish begins to move, retrieve the fly in such a way as to hop or drag it a short distance, then let it settle to the bottom. This will often be enough movement to attract the carp's attention but not spook it. If it all works as intended, the fish will advance to the fly and suck it in. Set the hook and hang on!

Carp fishing, while gaining more acceptance, is still not on many anglers' radar, so one of the advantages is that you normally will have very little competition. This fish is definitely worth being added to the freshwater species that you will want to pursue.

9

Walleye, Pike, and Musky

There is no substitute for fishing sense, and if a man doesn't have it, verily, he may cast like an angel and still use his creel largely to transport sandwiches and beer.

—Robert Traver, *Trout Madness*

I grabbed the bowline of my boat and pushed it off the trailer. After wrapping the line around a post on the dock at the boat ramp, I drove my Suburban up the ramp and parked car and trailer in the adjacent lot. I let my dog, Daisy, out of the car and she charged down to the dock as I walked. Already in the boat by the time I arrived, she had not been distracted by the sound of the waves rolling in on Lake Superior, which was just a short distance away, nor the sound of the many seagulls wheeling along the shoreline.

I launched the boat off the ramp, started up my small outboard, and began to motor up the river. Here at the mouth of the Tahquamenon River, along the Lake Superior shoreline, we were cruising through one of the wider parts of the river. After passing the campground, the shoreline had few homes and took on the feel of being in the Upper Peninsula; cedar, hemlock, and spruce trees were in abundance, along with the occasional white pine that towered above the others. Shoreline cover consisted of downed trees

A nice musky (in the bar phase) caught from one of Michigan's Upper Peninsula lakes.

131

The Lower Falls of the Tahquamenon River. The 17 miles from here downstream to the mouth of the river provide chances to catch walleye, pike, and musky.

Brightly colored flies are best used in tannic-colored waters like that of the Tahquamenon River.

along with beds of lily pads. This was my first time here, and there was a sense of anticipation as we continued upriver.

The Tahquamenon River's headwaters are in the northeastern portion of the Upper Peninsula of Michigan. Beginning in the boreal wetlands rich in cedar, hemlock, and spruce trees, the water is the color of a good Earl Grey tea. It flows almost 90 miles before emptying into Lake Superior. Here the outflow creates a tea-colored plume dissipating into the gin-clear waters of the largest of the Great Lakes. The upper reaches are considered a quality brook trout fishery that would have appealed to the likes of John Volker (aka Robert Traver), whose books *Trout Magic* and *Trout Madness* chronicled his fishing adventures in the Upper Peninsula. This is also Big Two Hearted River country, as that river mouth is on the west side of White Fish Point, while the Tahquamenon River enters Lake Superior on the east side.

The Tahquamenon River is noted for both its Upper Falls and Lower Falls. From the mouth of the river to the Lower Falls, a distance of 17 miles, the river can be nearly 200 yards wide in places with depths of up to 20 feet or more. The uniqueness of this river is that along with yellow perch and smallmouth bass, it has three other species—walleye, musky, and pike—in one watershed. Most of the lakes in the Upper Peninsula have perhaps two of these fish, but rarely all three.

Jason Haddix, a gifted Colorado fly tier, specializes in tying large flies for many species of fish. This box of Jason's patterns contains perhaps all the flies needed to fish for the walleye, pike, and musky. His flies are so beautifully and artistically tied that I sometimes find it difficult to fish them.

Both pike and musky are members of the pike family; the walleye is the largest member of the perch family. Even though they come from different families of fish, there are several reasons why I chose to put them together in the same chapter.

First, all three fish, as opposed to the others in this book, would be considered cool-water fish. They prefer slightly lower temperatures than the sunfish family for both spawning and feeding. Throughout the Great Lakes region, all three of these fish will also start spawning shortly after ice-out, long before the sunfishes spawn. They also share an overlapping and common area of distribution that stretches along the northern tier of states and into Canada. While there is some distribution in the central and southern states, they will usually not attain the same size as in the cooler waters to the north.

Additionally, there is a great similarity in fishing methods for all three. While each fish certainly has its unique tactical considerations, walleye, pike, and musky are most commonly fished for using sink-tip lines and streamer or baitfish patterns. Some of the same fly patterns can be fished for all three species.

Pike and musky are more commonly fished for with a fly than walleye, but walleye have qualities and their own set of challenges that make them worth pursuing.

ABOUT WALLEYE

For fly fishers, walleye is probably the species that is the least pursued, as well as the one that is the least written about. All the other species written about here have books dedicated to them; however, I am unaware of any fly-fishing books devoted to walleye. There is good reason for this to be true, as walleye are rarely obtainable using fly-fishing methods.

Walleye go through the same stages of pre-spawn, spawn, post-spawn, summer peak, and fall periods as all other warmwater and cool-water fishes. The characteristics of each of these stages will be similar as well. With walleye, pre-spawn happens so early that in the northern tier of states the ice can still be present, depending on the spring weather. Walleye will begin spawning early, often shortly after ice-out when water temperatures are as low as 38 degrees up to 50 degrees. Water temperatures that are in the mid-40s are preferred. Hard gravel and rock bottom with some current are needed for the nests. Walleye will look for these areas in up to 5 feet of water. Like trout eggs, walleye eggs need to be constantly aerated, so common places to find spawning sites are inlets with river currents, rocky shores, or mid-lake humps with good wave action. As with other fish, I choose not to fish over active spawning sites. If the lake also has a population of bass, this is also a good time to target them in their pre-spawn stage.

Walleye can be caught on flies in the right locations. This walleye was caught in a pond where the water depths did not exceed 10 feet.

When walleye enter the post-spawn and summer peak, it will become challenging to catch them with flies. Given the opportunity, they will always move off to deeper water. They mainly forage for baitfish, and if the baitfish are in deeper water, the walleye will follow, rarely returning to shallow water. Walleye prefer low-light conditions. As their name implies, they have very unusual eyes and are adapted to live and feed with very little light. Their eyes have a special reflective membrane that gathers a high percentage of the available light, which makes them very effective low-light predators. However, it is this same feature that sends them to the deeper water in the summer to escape the brighter light.

For people who fly fish, the catch-and-release ethic is deeply ingrained, especially if coming out of a trout-fishing tradition. This is true for me as well. However! I move forward here with trepidation, lest you think I am a heretic. Walleye are perhaps more noted as table fare than for their sporting qualities. I have family members who are dedicated walleye fishermen. I am sure they appreciate the sporting qualities of catching them, but if large enough, they will quickly dispatch them. Having perfected the art of battered, deep-fried fish dinners, which I have enjoyed, it is hard to argue the merits of catch-and-release fishing here. Scattered across much of northern Michigan, especially in the Upper Peninsula, are numerous small pubs. While burgers and beer are the main fare, most of them also offer a fish-and-chips basket with walleye as the main course. If, like me, you like fish, these places should be part of any trip to this region.

ABOUT PIKE

With coffee in hand, I really enjoy walking out to the end of our dock to watch the morning begin to unfold. I consider it fortunate if the loons are present and calling to one another. The sound is beautiful as it reverberates throughout the lake, in what is such an iconic voice of the northern Great Lakes region. About the only other sound in nature that I think rivals it is the bugling of elk on an early morning in the mountains of western Colorado.

Of course, if the loons are absent, there are the ever-present panfish that are always swimming around and under the dock. Our lake is spring-fed and, if left undisturbed by excessive boat traffic, incredibly clear. When at its clearest, the bottom of the lake is visible in water depths of 10 feet or more.

Our dock sits on a hard sand bottom that extends into the lake. Another dozen feet out from the end of the dock, the bottom changes color and drops off into the deeper water. At the edge of the color change is an underwater

These are the teeth and eyes of a predator. Even a smaller pike needs to be handled with care.

A nice pike that was caught (although not with a fly; note the rod and lure in his left hand) by my paternal grandfather, Ed Jacobs, pictured here with my dad, Roland Jacobs, circa 1926.

weed bed of coontail and pondweed that acts like a wall separating the shallow, hard sand bottom area from the deeper reaches of the lake. On this particular summer morning, the loons were absent, so I was watching the small panfish swimming around. Suddenly something shot out of the weed bed and disappeared under the dock. It was then that I noticed one of the panfish was motionless and beginning to float sideways. Out from under the dock came a pike leisurely swimming along. It casually swung around to eat the panfish head-first before swimming back into the darkness of the deeper water and weed bed.

It all happened much faster than I was able to write these sentences describing it. What amazed me about this event was, first, that a fish that large was that close and I had no idea it was there. Second was the explosive speed of its ambush. It was more an arrow-like flash of light than any perception of a fish. Until it came out from under the dock to collect its breakfast, I really had no idea what had happened. This is typical of pike, as they are shallow-water ambush predators. Their coloration makes them hard to locate, but fishing streamer patterns around shallow-water cover is always a good tactic for pursuing them.

Pike go through the same stages of pre-spawn, spawn, post-spawn, summer peak, and fall periods. Like walleye, they begin spawning shortly after the ice comes off the lake. As soon as the water temperatures rise above 40 degrees, they will begin spawning. Unlike the sunfishes, though, they do not sweep out nests. Instead, they will look for shallow-water areas, with vegetation, that will warm up the fastest. They then will spawn by broadcasting the eggs throughout the vegetation, and the males will fertilize the eggs. This includes a great deal of splashing and thrashing to ensure the eggs are fertilized and spread throughout. Once the spawning is complete, they will leave the area. When the eggs mature and hatch, the fry will be on their own.

As with all the other fishes described in this book, I will not fish for spawning pike. First of all, when pike are spawning, they are really focused on that activity. Also, they do not have the same territorial instincts concerning their nesting area that sunfishes do, so fishing for them at this time is not very productive.

Post-spawn and summer peak are more like one season for pike. They will begin feeding right after spawning. This post-spawn period will be some of the best fishing of the year, as water temperatures are still cool enough that the fish will remain in shallow water, and they will be aggressively feeding. In lakes with both pike and bass, it is not unusual to hook pike in the late spring and early summer in the shallow-water cover. As the summer progresses and the water temperatures begin increasing, the pike will move out to the deeper, cooler water to be replaced by the bass as the dominant shallow-water fish. Then in the fall as the water temperatures begin dropping, this process reverses. As you travel north, lakes will remain cooler longer into the summer. Here pike will be in the shallow water a correspondingly longer amount of time. As you get into Canada, where the water temperatures remain cool throughout the summer, pike will stay in shallow water year-round.

While bass do not hunt in schools, I have occasionally seen loose groups of bass cruising the shorelines hunting for baitfish. Not so with pike. I will seldom see them cruising. Instead, they are solitary hunters and almost always feed by ambushing the baitfish. It is the ability to rapidly accelerate in a straight line that makes them such an efficient predator. When I do see pike, they are normally stationary along or in some form of cover. Fishing baitfish patterns around this type of likely cover, which would conceal a pike's presence, is always a good strategy.

Unlike walleye, pike will strike surface flies. I have on more than one occasion hooked a pike on a topwater or diving bass bug while fishing for bass. Of course, the result depends on how deeply the pike takes the fly. If hooked in the corner or edge of the mouth, I may land it. If the fish devours the fly, it is then a very short fight ending in a cleanly cut leader.

Given cool-enough water temperatures, this fish prefers to hunt baitfish in shallow water, making it one of my favorite targets for the fly rod.

ABOUT MUSKY

I had spent the earlier part of the morning fishing in the main body of the lake. Having caught only a couple of the smaller musky, I also had a few follows from some very large fish. From the standpoint of musky fishing, this was not a bad start to spending some days in the Upper Peninsula of Michigan. This was one of my favorite UP musky lakes. At only 140 acres, it was small enough to be very comfortable in my boat. It was completely undeveloped but had a gravel boat ramp, making access fairly easy. Rarely did I ever share the lake with anyone other than the loons.

I fished around a point that opened into a system of three bays, each one successively smaller. There were narrow and somewhat shallow passages between each of the bays. The shoreline cover consisted of downed trees and lily pads, and these bays in particular had a lot of downed trees along the banks. I had been fishing a sink-tip line with a perch-colored articulated streamer pattern most of the morning. Since this water was much shallower

A tiger musky caught from a western reservoir. PHOTO BY FRANK WHISPELL AT WWW.FFFADVENTURES.COM

than the main body of the lake, I switched over to a lighter rod with a float-ing line and a feather-winged deer hair diver.

As I entered the middle bay of the three bays, I continued to cast off the left side of my boat to the wood cover along the bank. This bay was quite shallow in the middle, with a gray mud bottom about 2 feet deep. The deeper water was along each side of the bay. With downed trees and a few lily pads, there was ample cover creating prime locations for the fish.

A disturbance on the opposite side of the bay drew my attention. Charging across the shallow water, leaving a mud trail behind it, a musky was headed directly for my diver. As the fish drew closer, I slowed down my retrieve momentarily and then began to speed it up as the fish approached. I was hoping that by simulating an injured baitfish attempting to escape, I would elicit a strike. But as quickly as the fish had charged across the bay, it just as abruptly stopped. After a brief inspection of my fly, it casually turned and disappeared into the larger bay behind me.

This is both the frustrating and riveting aspect of musky fishing that capti-vates so many fly anglers. Along with carp, musky (or muskellunge) are among the most difficult freshwater fish to catch. However, when I fail with carp, it is normally because I have spooked them with my approach. When I fail with musky, it often remains a mystery. Although the musky is the largest member of the pike family and looks very similar to pike, they behave differently.

Musky go through the same stages of pre-spawn, spawn, post-spawn, sum-mer peak, and fall periods as all the other fishes described here. They will

Perch are a preferred forage fish for both pike and musky.

Casting to cover and edges is always a good practice.

spawn after pike, preferring warmer water. Once water temperatures climb from the high 40s into the 50-degree range, they will begin to search out areas to spawn. Musky are broadcast spawners like pike and will disperse their eggs into shallow-water vegetation. Also like pike, once the spawn is completed, they will leave the area and almost immediately begin to forage, providing a good opportunity to catch them. Fish may feed somewhat more aggressively during the post-spawn period but as the water warms, they settle into the summer peak period.

Rarely do you see musky just cruising along a shoreline. Instead, like pike, these fish are solitary hunters and prefer to station themselves in advantageous places in order to ambush baitfish. Also like pike, they have great forward acceleration.

Normally you will be fishing the structure and cover rather than sight-fishing. I have seen large musky resting in the shade under lily pads during the day, but rarely have I been able to tempt them to take a fly. They will usually just move to another location in the lily pad bed or swim out to deeper

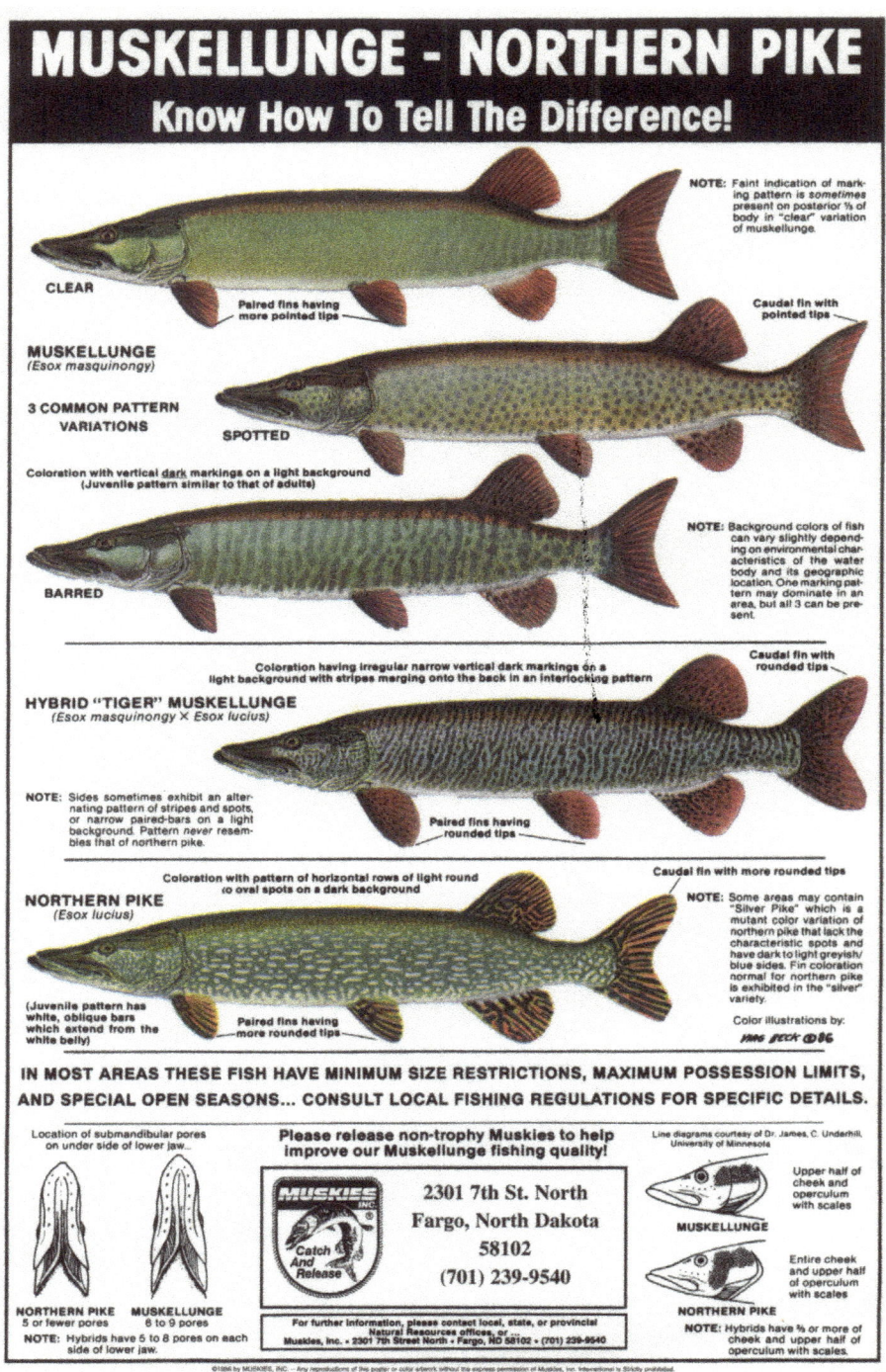

Muskie-pike identification chart. Reprinted with the permission of Muskies, Inc.

water. Like the pike off of my dock, when musky are looking to feed, they seem to settle inconspicuously into the cover, waiting for an unsuspecting baitfish to travel past.

The tiger musky is a musky-pike hybrid. They are sterile and sometimes used to control overpopulated panfish or other species. Tiger musky will occur naturally in waters where both pike and musky are present as they can cross-breed naturally. They can be found all across the Great Lakes region in states such as Minnesota, Wisconsin, and Michigan, as well as in Canada. They can also be found as far east as New Hampshire and Massachusetts, and as far south as Arkansas. Tiger muskies are especially popular in western reservoirs in Montana, Idaho, Utah, Colorado, New Mexico, and Wyoming. Fishing tactics that work for pike and musky will be just as effective with tiger musky.

TACKLE, FLIES, AND TACTICS FOR WALLEYE, PIKE, AND MUSKY

I have a variety of rods that range from 6-weights to 11-weights for this type of fishing. The 10-weight and 11-weight rods are reserved for pike and musky fishing only. The 11-weight musky rod is a heavy rod to fish all day. In fact, if fishing alone, I will always have a few lighter rods available so I can take a break from casting a rod this heavy. If fishing with another person, we will trade off, each of us casting and fishing for about 30 minutes before switching.

I have yet to crack the 50-inch mark for musky or pike. However, I have caught a number of both pike and musky that exceeded 40 inches using rods as light as a 6-weight. While the bigger rods are needed to cast big flies, I can cast most flies up to 8 inches long with lighter rods. If looking for a rod that would work for all three fish, probably a 9-foot 8-weight would be the one. With the right line, this rod can cast walleye-size flies easily, as well as handle moderate-size pike and musky flies. The size of the rod is dictated more by the size of the flies being cast, rather than the fish you are trying to catch.

As stated earlier, I only use large-arbor reels. All the reasons listed elsewhere also apply here. When considering a reel for a 10-weight or 11-weight line, a saltwater reel may be needed to accommodate the line. The largest freshwater reels will normally accommodate up to a 9-weight or 10-wieght line. Depending on the taper of the line, sometimes a 10-weight line may begin to max out the reel's capacity. For reels with 6-weight lines or above, I will use a 100-yard spool of 30-pound Dacron backing. A properly sized reel should accommodate 100 yards of backing plus the appropriate-size line.

Walleye will almost always be found at the bottom of the water column in whatever water you are fishing. This is why you need to be selective in finding fishing locations with depths less than 12 to 15 feet. Sometimes, during the low-light hours of early morning or late evening, when baitfish are present, they will return to shallow water to forage. Fishing in these conditions will

This box of assorted pike and musky flies includes articulated streamers, large waking and diving deer hair flies, and few large articulated deer hair poppers. All of these flies require a large 10- or 11-weight rod to cast them. When fishing a floating line, there are commercial leaders available that have bite tippets and clips for attaching flies. If fishing a sink tip, it is best to tie a shorter leader with a bite tippet. Tie a clip to the end of the bite tippet to make it easier to attach the fly. Also, pike and musky rods often have extended butts to facilitate doing a figure-eight as part of finishing up the retrieve.

sometimes be productive. The best conditions to fish for walleye are when they are present in shallow-water lakes, ponds, or rivers and do not have deep water, as they can be targeted with flies. If the water being fished is shallow enough to reach the bottom with a floating line and leader, use that setup. Most of the time, though, a sink-tip or sinking line will be needed. When considering the leader, the requirements for walleye are different than what is needed for pike and musky. Virtually all walleye fishing will entail using a sink-tip fly line or an added sink tip attached to a floating fly line. Normally, when using a sink tip, I will tie my own leaders using Maxima Chameleon for the first two sections, starting with a section of 20-pound test followed by a section of 15-pound test. I then use 12-pound-test fluorocarbon as the tip section, attaching the three pieces of material using a blood or barrel knot. This simple three-section leader should be only about 6 feet long. A commercial tapered leader could be used as well; however, keep the overall length of leader under 8 feet long. Suitable flies for walleye would include Clouser Minnows and synthetic-fiber baitfish patterns, as well as Woolly Buggers. Flies tied in white, black, and especially chartreuse are the most effective. Since these flies will be fished primarily in low-light conditions or in turbid water, the bright colors are more effective. Black is noticeable in these conditions as well. Small hair-jig patterns that can be retrieved near or on the bottom are also effective. I will always start with a chartreuse fly and change colors later if I am not getting any fish. Allow flies to get to the required depth, then strip in short amounts of line to jig or hop along the bottom. Slow retrieves are normally best.

Pike and musky both require leaders with bite tippets. I use bite tippets of either wire or heavy saltwater fluorocarbon. Wire bite tippet material normally comes in either 20-pound or 30-pound test. Hard saltwater fluorocarbon is available in sizes up to 100-pound test; however, 35-pound and 40-pound test is probably sufficient. When attaching wire bite tippet material to a leader, I use an Albright knot. For attaching fluorocarbon bite tippet, I use a slim beauty knot. I also attach a clip to the end of the bite tippet material, as this facilitates changing flies without having to retie knots. If you want to avoid the more complicated knots needed to attach bite tippets to leaders along with attaching the flies to the bite tippets, commercial bite tippet leaders are available. (For more information on tackle, see chapter 3, "Modern Tackle for Warmwater Fly Fishing.") Synthetic-fiber baitfish patterns are some of my favorite flies here. In addition to triggering strikes, these patterns are extremely durable. Color schemes for these flies can range from the subtle colors of the prevailing baitfish to bright colors such as chartreuse, yellow, or white. On the darker side, flies that are purple and black or all black can also be effective. Keep a small comb in your tackle bag so you can occasionally comb out the tangles. This will improve the movement of the fly in the water, giving it a better swimming action. Flies tied with rabbit are also effective as well as durable. Large topwater deer hair bass bugs and deer hair divers are effective as well. Be forewarned, though: a pike can quickly destroy a deer hair fly. However, I am willing to sacrifice a few deer hair flies to catch pike on the surface. For both pike and musky, I will begin by targeting the shallow water cover. Lily pads or reeds provide excellent cover for an ambushing fish. Start with a floating line and a diving fly. I use diver flies more often here than a topwater popping type of fly. When a pike or musky is waiting under the cover of lily pads, the diver fly will drop below the cover and is more readily seen. The diving action of these flies will often trigger an aggressive response. If divers prove unproductive, then add an appropriate sink tip to the line and fish a steamer baitfish pattern along the edges of the cover. When perch share the same water, they are one of the prime forage fish for musky. Streamers tied to represent perch are always a good place to start when selecting a fly. Fire tiger color schemes with a chartreuse back and orange belly are also effective, especially in tannic or turbid water. Chartreuse flies are effective for all three fish in this chapter. I have tied and used flies with just an articulated tail to good effect. However, articulated flies with three or more sections are effective for all three species, especially musky. The Flymen Fishing Company manufactures shanks for tying articulated flies. They come in a variety of shank

Releasing a musky in the Upper Peninsula of Michigan.

Hard-bottom beach areas provide the right set of circumstances to wade and fish the color break and drop-off.

lengths, so articulated flies can be tied in relatively small sizes of a few inches up to a foot.

I have primarily fished for musky in smaller lakes. A phenomenon that is common with all these smaller lakes is that musky will often follow larger flies, meaning flies that are 10 inches or longer, but will rarely eat them. All the musky that I have caught have come on flies that are 6 to 8 inches long. I am not sure why this is, but I suspect that, like with bass, smaller flies will elicit a more aggressive response from fish, whereas they tend to be more cautious when approaching larger flies. One of the tactics I have used, once I have fished the shoreline cover on small lakes, is to fish in bays with bottom weeds that do not reach the surface. Areas that are from 10 feet to 20 feet will often have weeds with a few feet of open water above the weed bed. Depending on wind conditions, I will allow the boat to simply drift over these areas. Using a slow sink tip, I will count the fly down as it sinks so it fishes just above the top of the weeds. Once you have determined the correct count to allow the fly to sink down so it is fishing just above the weeds, continue to use the same count to allow the fly to sink the same distance with each cast before beginning the retrieve. As the boat drifts, cast in a fan pattern around the boat. Musky will often be located in these weed beds and will feed on baitfish that swim overhead. Some of my largest muskies have come from these types of areas.

While it is true that all three of these fish inhabit waters that have a lot of development along the shores, I associate these three fish with wilder, more undeveloped lakes and rivers. This aspect of accessing these wilder, more remote areas is one of the draws to pursuing walleye, pike, and musky.

AFTERWORD

People often ask why I fish, and after seventy-odd years, I am beginning to understand.
I fish because of Beauty.
Everything about our sport . . . is beautiful. Its more than five centuries of manuscript and books and folios are beautiful. Its artifacts of rods and beautifully machined reels are beautiful. Its old wading staffs and split-willow creels, and the delicate artifice of its flies, are beautiful. Dressing such confections of fur, feathers and steel is beautiful, and our worktables are littered with gorgeous scraps of tragopan and golden pheasant and blue chattered and Coq de Leon. The best of sporting art is beautiful. . . . Our methods of seeking them are beautiful, and we find ourselves enthralled with the quicksilver poetry of the fish.
And in our contentious time of partisan hubris, selfishness, and outright mendacity, Beauty itself may prove the most endangered thing of all.

—Ernest Schwiebert (2005)

As I sit here at the beginning of creating this book, it is the middle of August 2020, although I have been "writing" it for the past few decades. This is the first year in my fishing life that I have not spent the summer at our cottage in Michigan. My family and I decided that in accordance with the CDC and Colorado Department of Health travel advisories, we would forgo our trip back to Michigan. Now our valley is smoke filled from the Pine Gulch Fire near Grand Junction, and we can see columns of smoke from where the Grizzly Creek Fire is burning east of Glenwood Springs.

The list of major events this year is quite astounding, and any one of them would be a cause for a memory that would mark the year in an unprecedented way. Starting in March, I had to quickly leave my classroom of students and begin the process of virtual learning because of the COVID-19 pandemic. Then on May 19, I watched as record flooding in Midland County, Michigan, caused both the Wixom Lake dam to fail and the Sanford Lake dam to be breached. Having grown up in Midland County, I have fished both of these lakes as well as the river connecting them. It was a surreal experience to watch the video of the dams and flooding, recognizing almost all the locations. The

149

My maternal grandfather, George English, at our cottage on Lake George, 1956.

future of both lakes is unknown, as the prohibitive cost of building new dams makes their replacement uncertain at best.

On top of this are so many other serious issues, that writing a fly-fishing book against this backdrop of events seems like a bit of a nugatory endeavor. However, it may also be the most important thing I could do. Fly fishing has been one of those quiet constants in my life. Along with fly tying, fishing, and generally just working on my fly tackle, the fly-fishing community and the many friendships I have made here are the great rewards of this sport.

My first time fly fishing, or even trout fishing for that matter, was to the Pere Marquette River. Equipped with an old telescoping 8-weight rod and some ridiculous flies, I really did not stand a chance of catching a trout. An older, experienced fly fisherman on the river took pity on me. He completely rebuilt my leader and gave me the fly that soon caught my first trout. From that early act of kindness and sharing to this day, my experience with the fly-fishing community has always been positive. Politics, religion, and all those things that divide us simply do not become part of our interaction. I guess we are all too "geeked out" over the latest fly-tying material or piece of tackle to worry about such things.

There is an additional aspect of warmwater fly fishing that I really enjoy: my dogs can join me in the boat when I go fishing. While I am often the only angler in the boat, I am never alone. If you are a dog lover like me, you understand. When bird hunters pull out the shotguns, their dogs immediately know what that means, and their excitement is palpable. For my dogs, that trigger is the sound of a fly reel. It is one of the real joys of life to share their companionship.

We may say of angling as Dr. Boteler said of strawberries: "Doubtless God could have made a better berry, but doubtless God never did, and so, If I might be judge, God never did make a more calm, quiet and innocent recreation than angling."

—Izaak Walton, *The Compleat Angler*

BIBLIOGRAPHY

This bibliography lists all the books and magazine articles directly mentioned in the text. Additionally included are the books, magazine articles, online resources, and DVDs used in researching the history of anglers, flies, and tactics needed to fly fish for warmwater fishes.

Bates, Joseph D., Jr. *Streamer Fly Fishing in Fresh and Salt Water*. New York: D. Van Nostrand Company, 1950.

———. *Streamer Fly Tying and Fishing*. Harrisburg, PA: Stackpole Company, 1950, 1966.

———. *Streamers and Bucktails*. New York: Alfred A. Knopf, 1979.

Bergman, Ray. *Fresh-Water Bass*. New York: Alfred A. Knopf, 1947.

———. *With Fly, Plug & Bait*. New York: William Morrow & Company, 1947.

Betts, John. "Bucktails." *Fly Tyer* 5, no. 1 (Spring 1999): 27–32, 53–56.

Bosanko, Dave. *Fish of Michigan Field Guide*. Cambridge, MA: Adventure Publications, 2007.

———. *Fish of the Midwest*. Cambridge, MA: Adventure Publications, 2016.

Boyle, Robert H., and Dave Whitlock. *The Fly Tyer's Almanac*. New York: Nick Lyons Books, 1975.

Brooks, Joe. *Bass Bug Fishing*. New York: A. S. Barnes and Company, 1947.

Chocklett, Blane. *Game Changer: Tying Flies That Look & Swim Like the Real Thing*. Boiling Springs, PA: Headwater Books, 2020.

Clouser, Bob. *Clouser's Flies*. Mechanicsburg, PA: Stackpole Books, 2006.

Cohen, Pat. *Super Bass Flies: How to Tie and Fish the Most Effective Imitations*. New York: Skyhorse Publishing, 2020.

Colla, Sheila, Leif Richardson, and Paul Williams. *Bumble Bees of the Eastern United States*. Washington, DC: USDA Forest Service and the Pollinator Partnership, National Fish and Wildlife Foundation, 2011. http://www.fs.fed.us/wildflowers/pollinators/documents/BumbleBeeGuide East2011.pdf.

Currier, Jeff. *Currier's Quick and Easy Guide to Warmwater Fly Fishing*. Jackson, WY: Self-published, 2002.

Deeter, Kirk. *The Orvis Guide to Fly Fishing for Carp: Tips and Tricks for the Determined Angler*. Bloomington, IN: Stonefly Press, 2013.

Dolan, Tom. *Sports Afield: Know Your Fish.* New York: Sports Afield, 1960.

Ellis. Jack. *The Sunfishes.* New York: Lyons & Buford, 1995.

England, Tim. "Banana Bug and Eel Bug." *Fly Tyer* 6, no. 2 (Summer 1983): 54–57.

———. "Gator Bug." *Fly Tyer* 6, no. 4 (Winter 1984): 48–49.

———. "Tim England." *Fly Rod & Reel* 7, no. 3 (July/October 1985): 22–25.

Gee, Lacey E., and Erwin D. Sias. *Practical Flies and Their Construction.* 1966.

Gordon, Sid W. *How to Fish from Top to Bottom.* Harrisburg, PA: Stackpole Books, 1955.

Greenberg, Josh. *Rivers of Sand: Fly Fishing Michigan and the Great Lakes Region.* Guilford, CT: Lyons Press, 2014.

Hannon, Doug. *Big Bass Magic.* Brainerd, MN: In-Fisherman, 1986.

———. *Field Guide for Bass Fishing.* Taylor City, NC: Atlantic Publishing Co., 1980.

Hauptman, Cliff. *The Fly Fisher's Guide to Warmwater Lakes.* New York: Lyons and Burford, 1995.

Henshall, Dr. James A. *Book of the Black Bass.* Cincinnati, OH: Robert Clarke & Co., 1881.

Hills, John Waller. *A History of Fly Fishing for Trout.* New York: Freshet Press, 1971.

Holschlag, Tim. *Smallmouth Fly Fishing: The Best Techniques, Flies and Destinations.* Minneapolis, MN: Smallmouth Angling Press, 2005.

Jacobs, Tim. *Tying and Fishing Deer Hair Flies.* Guilford, CT: Stackpole Books, 2018.

Jaworowski, Ed. *Perfecting the Cast.* Guilford, CT: Stackpole Books, 2020.

Jones, Keith A. *Knowing Bass: The Scientific Approach to Catching More Fish.* Guilford, CT: Lyons Press, 2002.

Junker, Patricia, with Sarah Burns. *Winslow Homer: Artist and Angler.* Fort Worth, TX: Amon Carter Museum; San Francisco: San Francisco Museum of Fine Arts, 2002.

Keith, Tom. *Fly Tying and Fishing for Bass and Panfish.* Portland, OR: Frank Amato Publications, 1989.

Kreh, Bernard "Lefty." *Lefty's Little Library of Fly Fishing: Advanced Fly Casting.* Birmingham, AL: Odysseus Editions, 1994.

———. *Lefty's Little Library of Fly Fishing: Fly Fishing for Bass.* Birmingham, AL: Odysseus Editions, 1993.

———. *Lefty's Little Library of Fly Fishing: Fly Fishing Techniques and Tactics.* Birmingham, AL: Odysseus Editions, 1991.

Krieger, Mel. *The Essence of Fly Casting.* San Francisco: Club Pacific, 1987.

Kustich, Rick. *Hunting Musky with a Fly.* Guilford, CT: Stackpole Books, 2017.

Law, Glenn. *Lefty's Little Library of Fly Fishing: A Concise History of Fly Fishing*. Birmingham, AL: Odysseus Editions, 1995.

Livingston, A. D. *Fly Rodding for Bass*. Philadelphia and New York: J. B. Lippincott Company, 1976.

Lyons, Nick. "The Bass Fly Revolution." Illustrated by Dave Whitlock. *Field & Stream* 95, no. 1 (May 1990): 54–58.

————. *The Quotable Fisherman*. New York: Skyhorse Publishing, 2010.

Maclean, Norman. *A River Runs Through It and Other Stories*. Chicago: University of Chicago Press, 1976.

Marbury, Mary Orvis. *Favorite Flies and Their Histories*. Boston: Houghton Mifflin & Co., 1892. Reprint, Secaucus, NJ: Wellfleet Press, 1988.

McClane, A. J. *McClane's Field Guide to Freshwater Fishes of North America*. New York: Holt, Rinehart and Winston, 1965.

McDonald, John. *The Origins of Angling*. New York: Lyons & Buford, 1957.

Meyer, Deke. *Tying Bass Flies: 12 of the Best*. Portland, OR: Frank Amato Publications, 1995.

Miyata, Ken. "Fishing Like a Predator." *Fly Fisherman* 16, no. 1 (1984): 50–53, 71–72.

Murray, Harry. *Fly Fishing for Smallmouth Bass*. New York: Lyons & Burford, 1989.

Pfeiffer, Boyd C. *Tying Warmwater Flies*. Iola, WI: Krause Publications, 2003.

Ryan, Will. *Northern Pike: A Complete Guide to Pike and Pike Fishing*. New York: Lyons Press, 2000.

————. *Smallmouth Strategies for the Fly Rod*. New York: Lyons & Burford, 1996.

Reynolds, Barry. *Mastering Pike on the Fly: Strategies and Techniques*. Boulder, CO: Johnson Books, 1993.

Reynolds, Barry, and Brad Befus. *Carp on the Fly: A Flyfishing Guide*. Boulder, CO: Johnson Books, 1997.

Reynolds, Barry, and John Berryman. *Beyond Trout: A Flyfishing Guide*. Boulder, CO: Johnson Books, 1995.

————. *Pike on the Fly*. Boulder, CO: Johnson Books, 1993.

Rosborough, E. H. "Polly." *Tying and Fishing the Fuzzy Nymphs*. Harrisburg, PA: Stackpole Books, 1978.

Schullery, Paul. *American Fly Fishing: A History*. New York: Nick Lyons Books, 1987.

————. "A Dreadful Scourge." *American Angler* 30, no. 4 (Summer 2007): 24–26.

Schultz, Ken. *Field Guide to Freshwater Fish*. Hoboken, NJ: John Wiley & Sons, 2004.

Schweitzer, Steve B. *Designing Poppers, Sliders & Divers*. Greeley, CO: Pixachrome Publishing, 2017.

Schwiebert, Ernest. *Trout*. London: Andre Deutsch Publishing, 1978.

Stewart, Dick, and Farrow Allen. *Flies for Bass & Panfish*. Intervale, NH: Northland Press, 1992.

Tapply, William G. *Bass Bug Fishing*. New York: Lyons Press, 1999.

———. "From Bobs to Bugs." *Warmwater Fly Fishing* 3, no. 1 (February/March 1999): 11–13.

———. "From Bobs to Bugs, Part II." *Warmwater Fly Fishing* 3, no. 2 (April/May 1999): 6–10.

———. "From Bobs to Bugs, Part III." *Warmwater Fly Fishing* 3, no. 3 (June/July 1999): 8–11.

———. "Tap's Bug." *Warmwater Fly Fishing* 2, no. 2 (April/May 1998): 22–25.

Tomes, Robert S. *Musky on the Fly*. Mill Creek, WA: Wild River Press, 2008.

Traver, Robert. *Trout Madness*. New York: St. Martins Press, 1979.

———. *Trout Magic*. New York: Crown Publishers, 1974.

Walton, Izaak, and Charles Cotton. *The Compleat Angler*. New York: Oxford University Press, 2014. First published in 1653.

Warren, Joe J. "The Science of Droguing." *Warmwater Fly Fishing* 2, no. 1 (February/March 1998): 76–77.

Whitlock, Dave. *L. L. Bean Fly Fishing for Bass Handbook*. New York: Lyons Press, 1988, 2000.

———. "Red Fox Squirrel Hair Nymph." *Fly Fisherman* 16, no. 5 (July 1985): 48–53.

———. *Tying and Fishing Dave Whitlock Originals*. Vol. 3, *Red Fox Squirrel Hair Nymphs*. DVD, 2004.

———. "Whitlock's Sheep Minnow Series." *Fly Fisherman* 31, no. 2 (February 2000): 40–45, 59.

Wilson, Terry, and Roxanne Wilson. *Bluegill: Fly Fishing & Flies*. Portland, OR: Frank Amato Publications, 1999.

———. *Largemouth Bass Fly-Fishing: Beyond the Basics*. Portland, OR: Frank Amato Publications, 2001.

Wulff, Joan Salvato. *Fly Casting Techniques*. New York: Nick Lyons Books, 1987.

———. *New Fly Casting Techniques*. Guilford, CT: Lyons Press, 2012.

Zimmerman, Jay. *The Best Bass Flies: How to Tie and Fish Them*. Guilford, CT: Stackpole Books, 2017.

FISHING RESOURCES

Frank's Fly-Fishing Adventures, Frank Whispell at www.fffadventures.com.

    Muskies, Inc., PO Box 1509, Waukesha, WI 53187-1509; https://muskies inc.org.

Sportsman Connection (1423 North Eight Street, Superior, WI 54880; https://scmaps.com). Sportsman Connection has lake map books for a number of states both in the Great Lakes region as well as some Southern states.

Bright Spot Maps (PO Box 1342, LaPorte, IN 46350). These books are no longer being published but can still be found by doing an internet search. They primarily cover Michigan, Indiana, and Ohio.

With an internet search you can find published resources such as fishing guides as well as lake and river map books for most states. Internet sites for state fishery agencies will often have information on many fishing locations as well. I also always use an internet resource such as Google Maps to look at possible fishing locations along with access roads to that location.

WRITING RESOURCES

"Writing is hard, even for authors who do it all the time. Less frequent practitioners . . . often get stuck in an awkward passage or find a muddle on their screens," wrote Roger Angell in his forward for *The Elements of Style*. I think many, if not all, of those who write something more than a grocery list would identify with this statement. The resources here are often what I turn to when trying to navigate the "muddle." My hope is that my writing is both readable and informative. Whether I achieve that goal is, of course, up to you, the reader.

Dreyer, Benjamin. *Dreyer's English: An Utterly Correct Guide to Clarity and Style*. New York: Penguin Random House, 2019.
This is absolutely the most readable writing source in my library. It's entertaining as well as informative.

Fowler, H. W. *A Dictionary of Modern English Usage*. New York: Oxford University Press, 2009. First published in 1926.
This is one of the more classic writing resources that I occasionally use.

Strunk, William, Jr., and E. B. White. *The Elements of Style*. 4th ed. Needham Heights, MA: Allyn & Bacon, 2000. Earlier editions 1959 and 1972 by Macmillan Publishing Co.
This is a small book at only 105 pages, including the index, yet it is considered one of the standard writing resources for the English language. William Strunk Jr. first wrote this book as a text to be used in his writing classes at Cornell University. E. B. White was a student of his in 1919. Thirty-eight years later, White took Strunk's book and added to it so as to create the book I now have. Almost two decades ago I inquired about writing for Fly Fisherman *magazine. They replied with a letter and a suggestion: Get a copy of Strunk and White's* Elements of Style. *While I have yet to make a contribution to the magazine, this small book has been read and reread. I am still working to digest the lessons it contains.*

INDEX

ABOUT THE AUTHOR

Tim Jacobs, a native of Michigan, is an angler, fly tier, photographer, writer, and German shorthair pointer aficionado from the Roaring Fork Valley, Colorado. He has guided in the Steamboat Springs area and the Roaring Fork Valley. Additionally, he was a Federation of Fly Fishers–certified casting instructor who taught the Sage Casting Clinics during the 1990s. He works as an instructor at the Michigan Youth Trout Camp each summer and is a casting instructor for numerous events, both in Colorado and Michigan. Tim has published a number of articles and authored the book *Tying and Fishing Deer Hair Flies* (Stackpole Books, 2018). He appears at numerous fly-fishing shows and fly shops as a demonstration fly tier for the Regal Vise Pro Staff and Nature's Spirit Pro Staff.

Printed in the USA
CPSIA information can be obtained
at www.ICGtesting.com
CBHW062338021124
16546CB00005B/1

9 780811 771122